ROUTLEDGE LIBRARY EDITIONS:
GEOLOGY

T0188255

Volume 13

GEOMORPHOLOGY:
PURE AND APPLIED

GEOMORPHOLOGY: PURE AND APPLIED

M.G. HART

Routledge
Taylor & Francis Group

LONDON AND NEW YORK

First published in 1986 by Allen & Unwin (Publishers) Ltd

This edition first published in 2020
by Routledge
2 Park Square, Milton Park, Abingdon, Oxon OX14 4RN

and by Routledge
52 Vanderbilt Avenue, New York, NY 10017

Routledge is an imprint of the Taylor & Francis Group, an informa business

© 1986 M.G. Hart

British Library Cataloguing in Publication Data
A catalogue record for this book is available from the British Library

ISBN: 978-0-367-18559-6 (Set)
ISBN: 978-0-429-19681-2 (Set) (ebk)
ISBN: 978-0-367-20743-4 (Volume 13) (hbk)
ISBN: 978-0-367-20749-6 (Volume 13) (pbk)
ISBN: 978-0-429-26325-5 (Volume 13) (ebk)

Publisher's Note
The publisher has gone to great lengths to ensure the quality of this reprint but points out that some imperfections in the original copies may be apparent.

Disclaimer
The publisher has made every effort to trace copyright holders and would welcome correspondence from those they have been unable to trace.

Geomorphology
PURE AND APPLIED

M.G. HART
Head of Geography, The King's School, Macclesfield

London
GEORGE ALLEN & UNWIN
Boston   Sydney

Allen & Unwin (Publishers) Ltd,
40 Museum Street, London WC1A 1LU, UK

Allen & Unwin (Publishers) Ltd,
Park Lane, Hemel Hempstead, Herts HP2 4TE, UK

Allen & Unwin Inc.,
8 Winchester Place, Winchester, Mass. 01890, USA

Allen & Unwin (Australia) Ltd,
8 Napier Street, North Sydney, NSW 2060, Australia

First published in 1986

British Library Cataloguing in Publication Data

Hart, M.G.
   Geomorphology: pure and applied.
1. Geomorphology
I. Title
551.4   GB401.5
ISBN 0-04-551087-3
ISBN 0-04-551088-1 Pbk

Library of Congress Cataloging in Publication Data

Hart, M.G. (Michael G.)
   Geomorphology, pure and applied.
Bibliography: p.
Includes index.
1. Geomorphology. I. Title.
GB401.5.H37   1985   551.4   85-18539
ISBN 0-04-551087-3 (U.S.)
ISBN 0-04-551088-1 (U.S.: pbk.)

Set in 10 on 12pt Palatino by Columns of Reading
and printed in Great Britain by Mackays of Chatham

Preface

This book can make some claim to originality. It is not an orthodox textbook on geomorphology. It does not provide the reader with the usual pot-pourri of facts about fluvial, glacial, periglacial, coastal, slope and desert landforms and processes. Instead it deals, in a systematic way, with those lines of enquiry and those concepts which cut right through the subject across the traditional divisions. It reveals the structure of geomorphology. Along the way I have also tried to indicate something of the history and nature of intellectual debate in the subject. Such a background knowledge is taken for granted in history and English literature, and it is perhaps a failing of geomorphology that our students remain rather ignorant of these matters. Some of the implications of modern knowledge and modern thinking are examined. Sometimes the implications are self-evident once you think of them. If the reader, in following through some lines of argument, finds himself thinking: 'Yes, this is interesting because all this is obvious except that it's never occurred to me before', then I am pleased because that is one of my objectives.

This is not a textbook specifically for any existing course in geomorphology, although it should prove useful for almost any course from the introductory level onwards. Some basic knowledge is assumed; all discussions begin there. No time is spent reviewing familiar ground, and no attempt is made to define terms that are well known or easily consulted. Hopefully, the reader will see that there is more to geomorphology than he or she thought, or be guided as to future reading, or see the subject through new eyes. My treatment of each topic is inevitably highly compressed. Whole textbooks have been written on matters that I deal with in one chapter or even less. For example, earthquakes and volcanoes are discussed in a few lines in Chapter 12; then in the Bibliography at the end of the book the reader can find half a dozen textbooks on just volcanoes and earthquakes.

This is, in part, a personal view of geomorphology, and it is intended to be. Virtually all academics could write such a book, and it would come out differently in each case. This book begins with the history of thought on geomorphology in the pre-Davisian era, and then considers the historical approach that dominated the subject during the first half of this century. Today's work is divided into two parts: modern pure geomorphology, and modern applied geomorphology. This is a classification that will be new to many readers, but it has long been in use in mathematics and indeed in other sciences, where theoretical knowledge is put to the test on some practical problem and then the results fed back into pure research. I think it is useful to look at geomorphology like that, and it is likely to become a standard classification of the subject.

M. G. Hart

Contents

Part III   Modern pure geomorphology

List of tables

Introduction

Most readers will be familiar with the type of Introduction written for the majority of textbooks. The bulk of the textbook deals with the chosen topic at the intended level of detail, but the Introduction spells out in very simple language what the author is trying to do, places the work in the context of geomorphology as a whole, and makes some appropriate comments about the general nature of the subject. In a way, this book is all Introduction, all context. It is a textbook about geomorphology, not a geomorphological textbook – a geomorphologist's bedside book. The body of knowledge that has accumulated in geomorphology is to be found in the many excellent books and articles on general geomorphology or some specific aspect of the subject. The reader is referred to those if it is part of that knowledge that he wants.

References, in fact, are an eternal problem. Obviously, one cannot quote them all. My intended readership consists of candidates preparing for university entrance exams, undergraduates, student teachers, A-level students and practising schoolteachers. All should find the book of some interest. Most, and especially A-level students and teachers, find it much easier to work with textbooks than magazine articles, so I have concentrated my references on the accessible books. I have mentioned magazine articles if they are classic works or if they make for smooth presentation of a line of thought. Too many references make for staccato reading, and I have made no attempt to be comprehensive. If someone's favourite work is missing, then it is an inevitable casualty of the style adopted.

The book opens with a review of intellectual debate in geomorphology in the pre-Davisian era. During this period the term 'geomorphology' did not exist, but nonetheless in retrospect we can see advances in science that form the origins of the subject. Chapter 1 traces the development of thought from catastrophism and its ultimate replacement by uniformitarianism and the Glacial Theory through to the pioneer work on process geomorphology done by American geologists towards the end of the 19th century.

Part II (Chs. 2–6) describes the historical approach to geomorphology that dominated the subject in Britain and elsewhere during the first half of the 20th century. This historical approach focuses on tracing the evolution or sequence of events that have led to the formation of a landscape. It appears first in the celebrated work of the American geomorphologist W. M. Davis, who synthesised geomorphology into a recognisable 'subject' for the first time using the unifying theme of the cycle of erosion. Over a period of approximately 40 years from about

1890 to 1930, Davis refined and in some respects modified his basic thesis, and at the same time other geomorphologists enthusiastically applied the cycle concept to specific environments such as arid, glacial and karst areas. In Britain, the line of thought suggested by the cycle was extended to form the basis of denudation chronology, an approach to geomorphology that dominated there from about 1930 to 1960. In recent years both the erosion cycle and denudation chronology have come in for considerable criticism, and the historical approach in general has fallen into disfavour. However, criticism of the cycle concept from French and German geomorphologists has led to the emergence of two important branches of the subject: climatic and structural geomorphology. Part II concludes with a consideration of Pleistocene geomorphology. One of the main aims of this aspect of the subject is to establish the sequence of events during the Ice Age, so this is the last, but important, survivor of the historical approach.

The remainder of the book is then devoted to modern geomorphology. I have chosen to classify the subject into two categories: pure geomorphology (Part III) and applied geomorphology (Part IV).

In identifying pure geomorphology as one of the two major branches of the subject today, we are focusing on the intellectual side of geomorphology and on pure research. It is, in a sense, work done for its own intrinsic academic value, although, as we shall see in Part IV, much of it can be, and is, being made use of in applied geomorphology. Similarly there is a return flow from the practicalities of solving actual problems back to pure research.

With just a few exceptions, applied geomorphology is a fairly new development, so traditionally geomorphology has been all pure. The term 'pure' has become appropriate only in the last 10 years or so with the appearance of applied studies.

It follows that the literature on pure geomorphology is truly voluminous, comprising the vast majority of work in the subject, although it is true that much of the work does have a practical application and that most of the textbooks also make some reference to applied studies. The main modern textbooks on geomorphology in general are those by Bloom (1969, 1978), Easterbrook (1969), Pitty (1971), Sparks (1972), Garner (1974), Ruhe (1975), Strahler (1973, 1975, 1976), Dury (1959), King (1976, 1980), Twidale (1976), Douglas (1977), Rice (1977), Thornes and Brunsden (1977), Gardner (1977), Embleton *et al.* (1978), McCullagh (1978), Small (1978), Cullingford *et al.* (1980), Knapp (1981), Melhorn and Flemal (1981), and Chorley *et al.* (1985). The level of treatment varies from A-level to final-year undergraduate. In addition there are the many textbooks concentrating on one of the traditional subdivisions of geomorphology. Thus on fluvial geomorphology one can cite Leopold *et al.* (1964), Morisawa (1968), Dury (1970), Chorley (1969c), Schumm (1972, 1977a, b), Weyman (1975), Gregory and Walling (1973), Gregory (1977), Smith and Stopp (1978), and Pitty (1979); on slopes, Brunsden (1971), Carson and Kirkby (1972), Young (1972), Schumm and Mosley (1973) and Finlayson and Statham (1980); the main works on glacial and periglacial geomorphology are those by Embleton and King (1975a, b), Flint (1971), Embleton (1972), Price and Sugden (1972), Andrews (1975), French (1976), Price (1972), Sugden and John (1976), and Washburn (1979); on coasts we have Steers (1964, 1980), Bird (1968), King (1972), Davies (1980), and Coates (1981a); on deserts the main texts are those by Cooke and Warren (1973), Doehring (1977), Goudie and Wilkinson (1977), Mabbutt (1977), McKee (1980), and Goudie and Watson (1981); and finally on karst, Jennings (1971), Sweeting (1972), Ford and Cullingford (1976) and Bögli (1980).

Most general textbooks on geomorphology discuss theoretical developments in the subject by taking these topics one by one – rivers, then slopes, then glacial features, then coasts, then deserts, and so on. Here, however, in keeping with the approach used in the rest of this book, the following chapters look at the basic structure of pure geomorphology – at those developments that cut right across these traditional subdivisions. We start with the study of land form, pass on to the processes responsible for them, then look at the properties of the materials of which they are comprised. Part III closes with a consideration of the methods used to analyse these three elements of pure geomorphology.

Applied geomorphology is made up of several distinct but related threads. First, there is man's effect on landforms and processes – man acting as a geomorphological agent. Secondly, there is the effect of geomorphology on man. Wherever man uses the land, he has to accommodate its relief, materials and water resources to his purposes. Sometimes geomorphological events occur with such intensity that they constitute a hazard to man. Thirdly, there is the extent to which geomorphology can contribute towards the solution of practical problems, towards the general needs of society. Presumably this must be regarded as the literal meaning of 'applied' geomorphology. It includes environmental management and the evaluation of resources, and it also includes the relevant techniques such as geomorphological mapping and landscape evaluation. These three threads are not separate. For example, man may recognise and evaluate a river flood as a natural hazard, manage the environment so as to minimise the hazard or its effects, and in so doing act as a geomorphological agent. Applied geomorphology is a coherent discipline.

Part V concludes the book with an assessment of geomorphology as a science and a summary of the various approaches and concepts that have dominated the subject over the years.

PART I

The history of geomorphology

1 The history of geomorphological debate in the pre-Davisian era

Introduction

The origins of geomorphology are obscure. The term itself was developed by geologists, probably WJ McGee and J. W. Powell, in the United States in the 1880s. Therefore, in the pre-Davisian era there was no science called geomorphology, but there were developments in geology and natural science that we can now recognise as early geomorphological thought. They are described in detail by Chorley *et al.* (1964).

Aristotle, the Greek philosopher and natural scientist, had several ideas that ring true today. For example, he conceived clearly the concept of a meteorological cycle, although curiously he applied it to the formation of water *within* the Earth. He observed that some streams resulted from the downward percolation of rainwater, although he could not believe that there was enough rain for all streams all the time. He also appreciated that streams carved the surface of the Earth and made the landscape.

Catastrophism

Such sound ideas played little part, however, in the mainstream of scientific thought during the Middle Ages (1066–1536). The source of water was thought to be beneath the earth, supplied from the oceans. Moreover, during these centuries biblical beliefs held sway: the basic idea was that the Earth had been shaped during a very short period of time, the six days of the Creation and the 40 days of Flood. Adherents to such a view were called 'catastrophists' or 'cataclysmysts'. There was even some degree of confidence over the timing of the event, for Dr John Lightfoot, Vice-Chancellor of Cambridge University, claimed in 1654 that 'Heaven and Earth . . . and clouds full of water and man were created by the Trinity on 26 October 4004 BC'. According to the catastrophists, therefore, the Earth was about 6000 years old.

There were dissenters, however; not all scientists were catastrophists.

For example, as early as the Renaissance (the 15th and early 16th centuries) engineers such as Leonardo da Vinci (1452–1519) and Bernard Palissy (1510–90) recognised the slow rate of operation of geomorphological processes. Palissy asserted that springs were fed by rain alone, but he did not convince natural scientists in general. But he also showed how it happened, giving a clear account of what we would now call the hydrological cycle. Leonardo was convinced that rivers erode their beds and thus excavate valleys. But his ideas remained unpublished until the end of the 18th century, they were not communicated to anyone, and they had little or no influence on the development of scientific thought. In fact, catastrophism remained important until the mid-19th century. In the meantime it is interesting to note that contemporary ignorance of the way in which rivers cut their valleys helps to explain the rather improbable river courses shown on medieval maps such as the 14th-century Mappa Mundi, now on display in Hereford Cathedral (Fig. 1.1). It is also perhaps significant that notions on gravity and its role were still rather vague. Newton's famous work came much later, in 1687.

We now enter a rather curious period, from the end of the Middle Ages to about the end of the 18th century, in which many valuable advances were made but which did not influence the consensus of scientific opinion regarding catastrophism. Most of the advances were made by practising engineers, and most of the engineers were French. In 1674 Pierre Perrault, working on the Seine basin, proved what Palissy had always maintained, that precipitation was adequate to sustain streamflow. In 1755 Euler founded the science of hydro-mechanics and stated its basic principles. In 1775 DeChézy proposed a formula for streamflow that is still in use today. During the 18th century, Lamblardie attempted to measure the drift of pebbles along the coast of the Pays de Caux in order to estimate the amount of dredging necessary to maintain small ports, and du Boys showed that a relationship existed between the size of particles carried by a river and the velocity of the current. The tradition of French engineers was continued early in the 19th century by Surell (1841) who, faced with the practical problems of the construction and maintenance of roads in the Alps, drew up a valid theory of fluvial erosion that included the concepts of base level and equilibrium profile. However, all these immensely practical findings by a small number of technicians had little impact on academic geomorphology. It is curious to reflect, in view of the importance attached today to the 'new' applied geomorphology, that the origins of the subject lie with practical engineers whose work was not embodied into the assemblage of knowledge at the time.

Meanwhile, also during the 18th century, some natural scientists were developing thoroughly the idea of stream sculpture. For example,

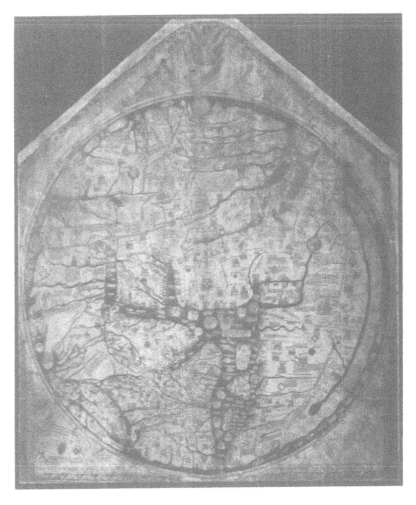

Figure 1.1 The 14th-century Mappa Mundi in Hereford Cathedral (from Brunsden & Doornkamp 1977).

Guettard's work in the Paris Basin and the Massif Central and his paper on the degradation of mountains dealt with the erosive ability of running water, and Desmarest traced the history and development of a landscape in the Auvergne, showing that the valleys were formed by the streams that still occupy them. Then, towards the end of the 18th century, we see academic geomorphology being born out of geology. De Saussure, for example, working in the Alps, saw that the valleys and drainage systems of the region were very closely related to each other, and he also saw fossil shells on mountain summits and doubted the concept of crustal stability. We can say that academic geomorphology derived from geology because de Saussure and other early geologists were the first of a long line of geologists who appear in the story shortly: Hutton, Lyell, Powell, Gilbert, even Davis himself.

However, that comes later. For the moment we stand at the end of the 18th century, and scientific opinion in general was still dominated by catastrophism. Indeed, the idea was still being verified and developed. In 1761 Alexander Catcott had described an experimental scale model in which he simulated the erosive potential of the Flood. So in 1800 it was not generally recognised that rivers cut valleys. Indeed, the very concept of the valley was a confusing one. Every depression was labelled as such, and for some even the Atlantic was a valley. Valleys pre-dated the rivers in them, so streams flowed in valleys because the valleys were there already. (In view of the fact that such notions were soon to be dispelled, it is interesting to consider that today that would be seen as a very reasonable view of present-day misfit streams flowing in oversize valleys or present-day streams flowing in glacial troughs.) We must also be clear that at the end of the 18th century there was no clear distinction between igneous and sedimentary rocks, geological time was still measured in thousands of years, and the erosional capacity of the sea and glaciers was appreciated no better than that of rivers.

Forty years later, by 1840, catastrophism had been refuted. The overthrow of this long-standing belief came in two stages: the appearance of the concept of uniformitarianism, and the rise of the Glacial Theory. In retrospect it can be seen as nothing less than a revolution in scientific thought.

Uniformitarianism

Uniformitarianism, often summarised by the little dictum 'the present is the key to the past', says that the processes and natural laws which existed in geological time are basically the same as the ones that may be observed in the landscape today. It follows that landforms have been formed by present-day processes operating slowly over long periods: the exact antithesis of catastrophism.

Uniformitarianism was initiated and established by Hutton (1726–97), a Scottish geologist, in his *Theory of the Earth* published in 1788. The concept was later persuasively argued and formally presented to the scientific community by Lyell (1797–1875) also a Scot and also a geologist (Fig. 1.2), in the highly successful *Principles of geology* (1830), which went through several editions. As geologists, they were aware of the unconformities in the English coalfields and they realised that mountains had been formed and levelled before the deposition of later strata. Lyell went on to describe precisely how processes disintegrate rocks, remove their debris by running water, destroy mountains in the long run, and lay down sediments from which the next generation of mountains will be built. There was also a third advocate of uniformitarianism, Playfair (Fig. 1.2) who published his *Illustrations of the Huttonian theory of the Earth* in 1802. In it are the main outlines of the concept of the drainage network, including the hierarchy of streams and a statement of 'Playfair's Law' which says that in areas of uniform bedrock and structure that have been subject to river erosion for a long time, valleys are proportional in size to the streams they contain, and stream junctions in these valleys are accordant.

At much the same time as the concept of uniformitarianism was being developed, it was becoming clear that the Earth was much older than the 6000 years proposed by the catastrophists. Indeed, the two ideas go together. The development of thought between about 1750 and 1850, during which the geological timescale was completely revised, is described in detail by Toulmin and Goodfield (1965). During the 1770s there were various attempts by scientists, notably that of Buffon, to apply Newton's new laws of physics to the problem of the age of the Earth. As it happens, Buffon totally underestimated the impact of radiant heat from the Sun on the rate of cooling of the Earth, and so his estimate of 168 000 years was a long way from the mark. Nonetheless, such ideas pointed the way forward. Hutton made it clear that the slow operation of geological processes required long time-periods, but he observed that geology proved nothing about the origin of the Earth ('. . . no vestige of a beginning . . .'). William Smith, the so-called father of stratigraphy, established the twin principles of superposition of strata and correlation of strata by fossils and thus further demonstrated the antiquity of the Earth. Lyell finally confirmed that geological time was more or less unlimited, although no

Figure 1.2 (left to right) Charles Lyell, John Playfair and Louis Agassiz (from Brunsden & Doornkamp 1977).

actual numbers were mentioned. Toulmin and Goodfield speak of '. . . Lyell's final breaching of the time-barrier . . .' (ibid., p. 171). The importance of this demonstration by scientists that the Earth is very old must not be underestimated: geomorphology could not have developed without it. Of course, the story goes on. Darwin's (1859) *Origin of species* implied long time-periods for man's existence, let alone the Earth's (and also, incidentally, with his theory of evolution, set the stage for Davis's evolutionary concepts on landforms), and modern radioactive dating techniques now show the Earth to be about 4500 million years old.

However, that is to digress from early 19th-century thought on uniformitarianism. It might be thought that the concept, thus argued by three prominent men (Hutton, Playfair and Lyell), would have forced the scientific community to discard catastrophism. But it didn't. For example, as late as 1815 the Niagara gorge was stated to have been formed by a dreadful catastrophe, and well into the 19th century many geologists were ascribing gorges in general to folding and faulting rather than to stream erosion. There were reasons why uniformitarianism was not accepted immediately: there were dry valleys in southern England with no rivers, in the Alps there were many examples of hanging valleys whose confluences with the main valleys were obviously not accordant, and there were lakes in many Alpine valleys. Another argument used against uniformitarianism was the existence of erratics and other superficial deposits that were supposed to have been distributed by so-called 'waves of translation'. Thus, although looking back we can now see that uniformitarianism 'replaced' catastrophism, the final death knell of catastrophism was sounded only several years later by the rise of a quite different branch of geomorphology – glaciation.

The Glacial Theory

Descriptions of glaciers can be found in 11th-century Icelandic literature, but the fact that they move does not appear to have been noticed, or at any rate recorded, until about 500 years later. In 1723 Scheuchzer put forward a theory of glacier movement: he suggested that water entered crevasses and froze, causing the ice to move downhill. By 1751, in the work of Altmann, we can see that it was recognised that gravity was the cause of ice motion. Later in the same century, de Saussure understood the formation, movement and some of the effects of glaciers, especially their ability to transport large boulders. By 1815, in the writings of Perraudin, a Swiss guide, the idea that the ice cover had formerly been more extensive comes in. De

Charpentier (1835) credited the poet Goethe with the discovery of the Ice Age around 1830, and the term 'ice age' was coined by Schimper in 1837.

It often happens in science that an idea which has been around for some time gets firmly accepted when a well known scientist puts his name to it. This is what happened with the Glacial Theory – the theory that ice can mould landscape and that formerly the ice was much more extensive. The scientist was Louis Agassiz (Fig. 1.2). He was convinced of the validity of the theory by de Charpentier and although he did not add much to the theory himself, his well known name gave greater weight to the new views. He explained his opinions in 1840 under the title *Études sur les glaciers* (Agassiz 1840a). He visited Britain in 1840 and concluded that the various upland areas he saw had all been under ice. In particular, he was impressed by the evidence afforded by polished and scratched rock surfaces, by erratics, and by superficial deposits such as sand and gravel, some of which contain erratics (Agassiz 1840b). Agassiz converted Lyell, Buckland and others to the Glacial Theory. Buckland had, in 1823, led what one might regard as a last rearguard action by catastrophists, and explained the existence of erratics and other superficial deposits as a result of the Flood, an idea that led to the introduction of the term 'drift'. However, Agassiz convinced him that the Glacial Theory could explain the origin of drift. In 1847 Agassiz left for North America where he again stimulated interest in the Theory. The relevance of his work in Britain was that, by explaining features such as hanging valleys and erratics, which had been stumbling blocks to the acceptance of uniformitarianism, he had freed geomorphology from the burden of the Flood. The sculpturing power of both glacial and fluvial processes was recognised, and catastrophism could finally be laid to rest.

But not quite. There was an unexpected delay in the complete acceptance of the Glacial Theory. In 1840, for example, although a great weight of evidence was presented to the members of the Geological Society in London by Agassiz, Buckland and Lyell, there was still considerable hesitation in accepting the implications of the Theory. This is because there is a fundamental conflict between accepting an ice age and being a uniformitarian. By the mid-19th century even Lyell himself had reverted to a form of the Flood Theory to explain the origin of erratics. They were supposedly deposited by icebergs floating in the sea, and the view was widely held at the time. Lyell had seen icebergs with boulders on them while on a voyage to North America and, as the present was supposedly the key to the past, he thought this was a better explanation for drift than invoking a cataclysmic extension of the ice caps. In 1860 Ramsay disproved this idea and thus increased support for the Glacial Theory in Britain, but one could still read about the iceberg theory in the literature of the 1880s.

However, during the second half of the 19th century Britain gradually disembarrassed itself of the legacy of the Flood. At this time there were two other significant developments that encouraged acceptance of the importance of erosion by streams. First, colonial geologists were seeing new landscapes, such as Blandford's studies in Ethiopia, which convinced them that water worked. Secondly, the 'old guard' in the Geological Survey in Britain had either died or retired, and were replaced by a group of scientists who were encouraged, as part of a new policy insisted on by Murchison, to include a piece on scenery in all regional monographs. They accepted the importance of fluvial activity, as can be seen, for example, in the approach used by Geikie (1865) in his textbook on physical geography.

This is the time to leave the situation in Britain. For, as we come into the last quarter of the 19th century, progress in geomorphology begins to accelerate. The scene now switches to the western United States, and the subject starts to develop along entirely different lines.

The beginnings of process geomorphology

During the late 19th century there was a series of exploratory surveys of the United States Geographical and Geological Survey, work associated with the names Powell, Dutton, McGee and, above all, Gilbert (Fig. 1.3). They made direct observations of the landforms, structures and processes in the plateaux, mountains and deserts of the west. These were areas where simple structures were clearly visible

Figure 1.3 G. K. Gilbert in Colorado in 1894 (from Brunsden & Doornkamp 1977).

beneath scanty vegetation. The importance of the relationships between landform and process, and between landform and structure, were clear to see. Most of the work was concerned with slope form, process and evolution, and with fluvial processes. Many fundamental laws of river action were demonstrated, and many geological terms and ideas in use today were introduced. For example, Powell formalised the base-level concept proposed by Surell, and he described drainage basin systems, introducing the terms 'consequent' and 'antecedent'. His main publication came in 1875. Gilbert's classic work is contained largely in *Geology of the Henry Mountains* (1877) and *The transportation of debris by running water* (1914). His main concern was the analysis of the erosional and transportational powers of rivers. He conducted detailed studies of both natural and laboratory streams. He appreciated the existence of equilibrium and balance in fluvial systems, and he developed general principles. It is not surprising if all this strikes a familiar chord with the modern geomorphologist, for this American work represents modern process-form geomorphology in both content and approach. Gilbert in particular was a veritable pioneer of process studies.

There were several reasons why these American geologists were in a position to make such important advances.

(a) Compared to the situation in Britain, they were relatively unfettered by Establishment dogma.

(b) In the Geological Survey there was a critical mass of people who could bounce ideas off each other.

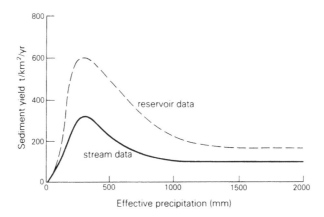

Figure 1.4 The relationship between effective precipitation and sediment yield obtained for the United States by W. B. Langbein and S. A. Schumm in 1958 (from Gregory & Walling 1973).

(c) In the Colorado Canyon there was evidence of very severe fluvial erosion because semi-arid areas have some of the highest rates of erosion in the world. This is something which came to light over 50 years later when Langbein and Schumm (1958) demonstrated that the relationship between effective precipitation and sediment yield peaks at a precipitation value typical of the semi-arid environment (Fig. 1.4).

(d) In the walls of the Grand Canyon there are enormous unconformities that were one of the prime sources of inspiration with regard to planation surfaces and base levelling.

Conclusion

However, the work just described made little immediate impact. It might well be thought that the strong lead given by Gilbert and his co-workers would have led on directly to the process geomorphology with which we are so familiar today, but in fact this is not what happened. This type of geomorphology was destined not to play a major part in the subject for many years, at least in Britain. For at this point there strode onto the stage a man whose ideas and teachings diverted attention away from process studies and dominated geomorphological thought for over 50 years: William Morris Davis. The man is nothing if not controversial. For many geomorphologists he is the key figure in the subject; for others, he merely delayed its development for half a century.

PART II

The historical approach in geomorphology

2 William Morris Davis: the geographical cycle of erosion

Introduction

Much has been written about Davis's cycle of erosion. Description and comment can be found in most general textbooks on geomorphology. Much is also known about the life, as well as the work, of Davis, as described by Chorley *et al.* (1973) in a book rich with often amusing anecdotal and personal detail. His work is also reviewed by King and Schumm (1980). Here it is proposed to review Davis's general approach to landform study and, in view of the very limited part played by that approach in modern pure and applied geomorphology, to summarise the ways in which most of today's geomorphologists would criticise Davis's work. Then, in the next few chapters, some further ramifications of that work will be discussed.

It is important to begin, however, by putting Davis's contribution to geomorphology into context. Of the various disciplines of the natural sciences, geomorphology was one of the last to appear. In a long series of influential publications between 1880 and 1938 – Davis was a prolific and productive writer – he synthesised the scattered elements of geomorphology and made it into a coherent subject for the first time. In retrospect we can see that the cycle of erosion is a geographical model and an early paradigm in geomorphology, a point discussed in more detail in Chapter 15. But, of course, Davis did not see it in those terms himself. His aim was to effect a general economy of description, to enable the trained reader to understand the descriptions of the trained observer of landforms, as both would speak the same language and have the same concepts. To this extent he succeeded magnificently. His work has proved to be a simple and effective vehicle for teaching and it has had a profound effect on the development of geomorphology in the English-speaking world, firing the imaginations of generations of students and accounting in no small measure for the popularity of the subject as a field of study.

The popular image of Davis is one of a man with rather entrenched views on just one topic, the erosion cycle. In view of that, two points must be made. First, he had great breadth of interest. For example,

Figure 2.1 William Morris Davis (from King & Schumm 1980).

within geomorphology he busied himself with glacial, coastal, arid and volcanic landforms in addition to the cycle. More generally, he was fascinated by the power of words. He was a student of the English language. We can put some of his success down to this. He knew how to write to make it sound convincing. Secondly, he changed his views considerably later in life, particularly on the universal applicability of the normal cycle. For example, he admitted the need for separate cycles in arid and glacial areas, and in 1930 he wrote that 'on shifting residence from one side of the continent to another, a geologist must learn his alphabet over again in an order appropriate to his new surroundings'.

The cycle of erosion

Detailed accounts of Davis's cycle of erosion abound in geomorphological textbooks, and this will be very familiar ground to most readers, so let us be brief here. Davis published his basic thesis on the cycle in 1899, considered some complications associated with the cycle in 1904, brought his main ideas together in a single work in 1909, and went on modifying and extending his views in a number of ways for the next 30 years.

He coined the familiar phrase: 'landforms are a function of structure, process and stage'. That is a very profound idea; if we define each of the three terms generously enough, it forms a sound basis for study. Certainly he adopted what we would consider a very wide definition of the term 'structure'. He did not mean just faults and folds. He embraced lithology and structure, rock properties at all scales. Davis, however, went on to concentrate (although not exclusively) on stage. He concentrated on the sequence of events or the origin and evolution of landforms. It is the genetic approach. It is the historical approach, the one that was taken up by the denudation chronologists a few years later.

Davis organised his evolutionary sequences of events into the concept of a cycle. The term 'cycle' had been introduced by Lawson in 1894, but Davis was to use it as his central theme. Using a simple descriptive, interpretative, deductive approach, and using simplifying assumptions (some of which he knew not to be justifiable in nature) to make the exposition easier, he explained how the cycle operated. An initial flat or nearly flat landscape is rapidly uplifted. Erosion then proceeds under conditions of prolonged tectonic stability and the landscape passes through the stages of youth, maturity and old age, each stage having distinctive and recognisable characteristics. The end product, the peneplain, is the same as the initial land surface – a flat or nearly flat landscape – and thus a cycle of events has truly been run, since renewed rapid uplift would trigger-off the same sequence of events again.

The processes eroding the landscape were grouped together under one heading, 'normal'. In some of his writings in German he specifically discussed the situation where uplift is not rapid, so that uplift and erosion occur together. The concept of base level is crucial in Davis's thesis, and he examined the effect of repeated minor uplift in detail, arguing that many landscapes are polycyclic in origin. He was also aware that the cycle might not be able to run its full course because of climatic change. He viewed glaciation, aridity and indeed vulcanicity as interruptions of the cycle and called them 'accidents', and he was even convinced of the need to formulate a separate cycle for arid regions.

Consistent with his historical approach, Davis had a long-term view of the landscape. He needed and used geological timescales – millions of years. Perhaps that is why his use of the term 'grade' has proved to be so confusing. By grade he meant a *short-term* balance between erosion and deposition, and between work to be done and the ability to do work. Streams were graded and yet landscapes were worn down from mountains to peneplains. It was a contradiction Davis was never able to resolve.

Criticisms of Davis's work

Although some schoolchildren, perhaps regrettably, are still taught geomorphology using the Davisian approach, it is quite possible to teach without any reference to the man or his ideas, and indeed modern geomorphology contains little or no Davisian thinking. This is because Davis's work has been criticised, some would say not surprisingly in view of its antiquity. The main criticisms can now be summarised as follows.

(a) In over-emphasizing stage and to a lesser extent structure, Davis did less than justice to his own dictum about landforms being a function of structure, process and stage. The criticism is not that this is intellectually untidy, but that the historical approach persisted as the dominant theme in British geomorphology for many years, until about 1960. This could hardly be said to be Davis's fault, but to many process geomorphologists, recalling Gilbert's perceptive work, Davis was responsible for delaying the development of geomorphology for over 50 years. The implications of all this are interesting because one of the reasons for the rise of modern process geomorphology and modern structural geomorphology is a deliberate attempt to redress the balance.

(b) It follows from what has been said above that Davis's greatest omission was the study, both in the field and in theory, of the detailed mechanics and nature of present-day processes. To some extent this reflects the descriptive and non-quantitative nature of Davis's work, which in themselves are important failings. The neglect of biological processes was complete. Davis's landscapes look like deserts: not a plant, not a tree. This is surprising since Davis based his work on the well vegetated temperate midlatitudes, and the neglect of biological processes make his choice of the adjective 'geographical' for his cycle seem singularly inappropriate.

Another criticism regarding processes is Davis's use of the term 'normal' to describe them. Davis did not explicitly define 'normal', but by implication he meant the assemblage of processes dominating the temperate landscapes of North America and Europe where he worked, and especially the action of running water. Now it is not unusual for geomorphologists to regard the landscapes with which they are familiar as normal. For example, the views of Walter Penck and Lester King are largely the products of the environments that they knew. Much more recently, Davies (1980), in the preface to his book on coastal geomorphology, writes '. . . it is part of my thesis that the development of thought on coastal processes and forms has been strongly influenced by the location of authoritative workers . . .' (ibid., p. v). Nevertheless, in view of the proportionately quite small extent of the humid temperate mid-latitude landscapes, one must question whether Davis's cycle and the regions to which it might apply are 'normal'. Moreover, there is a second sense in which the use of the term 'normal' is inappropriate, a temporal one. The mid-latitude landscapes are just emerging from an orogenic period which, in various places, folded, faulted and uplifted the surface. They are also just emerging from – or, to be more exact, are almost certainly in an interglacial stage of – the Ice Age which has left the imprint of glacial and periglacial processes, and which leaves a current legacy in the form of isostatic readjustment. Sea level, Davis's ultimate base level, has been roughly at its present level for only about 3000 years. Today, for the first time in a geological sense, man himself acts as a geomorphic agent. For Davis's type region, indeed for the world in general, there is nothing 'normal' about the present day. Let us not beat about the bush: in the context of conditions that have occurred in the geological past, present conditions are not normal, they are unique. A final reservation about Davis's treatment of process is his comparative neglect of depositional processes and landforms. This is faithfully reflected in the title – the cycle of *erosion*. This has been a traditional weakness of geomorphology over the years, although whether this is because of Davis's work or in spite of it is difficult to say. Whatever the reason, most general textbooks on geomorphology do less than justice to depositional processes and landforms, a situation only recently remedied, as discussed in Chapter 8.

(c) The initial assumption of rapid uplift is unsatisfactory. It is true that Davis was aware of this and discussed the implications. It is also true that contemporaneous erosion would almost certainly not match the rate of uplift. Nevertheless, the implications of erosion acting on a landmass as it is being uplifted slowly or intermittently are such as to cast doubt on whether the cycle would run the course that Davis described. In effect this is the loophole fastened onto by Penck, one of Davis's early critics, for, as will be seen later (Ch. 5), he considered landscape evolution to be controlled primarily by rate of uplift. Having said all this, however, it is interesting to reflect that the theme of Schumm's (1963) article on diastrophism is that the assumption of very rapid uplift is not necessarily far off the mark if one compares rates of uplift with rates of erosion. Some areas experience extremely high rates of uplift, and mean rates of orogeny could well be about eight times greater than maximum rates of denudation.

(d) It is doubtful whether the youth–maturity–old age sequence and the landforms associated with each stage actually occur. The problem is that the only *proof* of the cycle is its own logical and internal cohesion. The theory is not testable except against its own assumptions and conclusions. If in reality we observe a landscape with the features of, say, maturity, it is not possible to prove that it is mature because we cannot prove that it has been derived from a youthful landscape. There are some specific points to be made here. First, some landscapes have features of both maturity and old age. Secondly, some landscapes do not fit into any of the three categories. Landscapes exhibit an infinite gradation of form, not three types. Thirdly, it is doubtful whether surfaces flat enough to be called peneplains can ever be created by slope decline, as the capacity for work in the landscape becomes very low in the later stages. Fourthly, there is no evidence that slopes evolve in the way Davis suggested, and none, for example, that steeper slopes are younger than gentle ones. In fact, as a model for slope form and evolution, the cycle has proved misleading. Furthermore, the two main alternative models of slope evolution – those of Penck and King – have also produced an enormous amount of sterile and rather unedifying debate within the subject and also play little part in slope geomorphology today.

(e) Environmental change during the Pleistocene will prevent the cycle running its full course. Davis gave no precise figures regarding the time required for a cycle to do so, but on the basis of known denudation rates it can hardly be less than 10^6 years, and it could well be much more, perhaps even 10^8 years. Pleistocene climatic change is extremely complex, but oscillations have a period of at most 10^5 years. So obviously environmental change is going to intervene. Now Davis was aware of that, and admitted the existence of polycyclic landscapes, but that is not quite the point. The point is that environmental change means periodic change of *process*, so landforms will be polygenetic, not just

20

polycyclic. It seems certain that the peneplain would never be allowed to form, so the cycle, in the true sense of the term, would never be completed, and it is not surprising that nowhere on the Earth's surface is there a peneplain conceded by everyone to have been formed in the way Davis suggested.

(f) The concept of grade, as defined by Davis, has proved elusive. Discussion of this point has itself stimulated a massive and largely sterile literature. As explained earlier, it is difficult to envisage graded rivers and graded hillslopes co-existing with a long-term degrading of the landscape. The association of grade with maturity and with smoothly concave river long-profiles has also been unhelpful. The way out of these problems, incidentally, was shown much later by Schumm and Lichty (1965). The crucial point made by them is that any statement about grade needs to be accompanied by a statement of timescale. In the long run, the surface is worn down by the forces of denudation, so over periods of millions of years the landscape is not graded. But this might not be at all obvious over a period of, say, 10 years, when erosion and deposition do appear to balance. This is the sort of timescale Davis had in mind. Yet, at the same time, a river will scour its bed in a matter of days during every flood and fill it again afterwards, so the balance is a changing one – a dynamic equilibrium.

(g) The cycle of erosion has a deceptive simplicity which has tended to have a blinkering effect on its adherents. They then tend to find it hard to view landscapes in terms other than those proposed by Davis. To this must be added the point made earlier that Davis was a fluent, lucid and powerful writer. The easy style and the internal coherence of the argument combine to make the cycle seem more convincing than it really is.

(h) Davis never measured form. Impressions of slope form are notoriously inaccurate and misleading.

(i) This final criticism is more profound than all the others. It deals with Davis's basic method. His work is an example of the deductive approach. This means basing an argument on certain assumptions, and proceeding or inferring from the general to the particular. This is unscientific. Scientific method is the inductive approach, which builds on experimental evidence from particular instances to produce general inferences. This is the source of many specific errors made by Davis. For example, meandering is associated not with just maturity and old age. Meandering rivers also occur in landscapes that are otherwise youthful. Also, to take a more parochial example, Davis incorrectly attributed underfit streams on the Cotswold back-slope to river capture (1895, 1899). He even recognised one of the reasons why this cannot be so: the

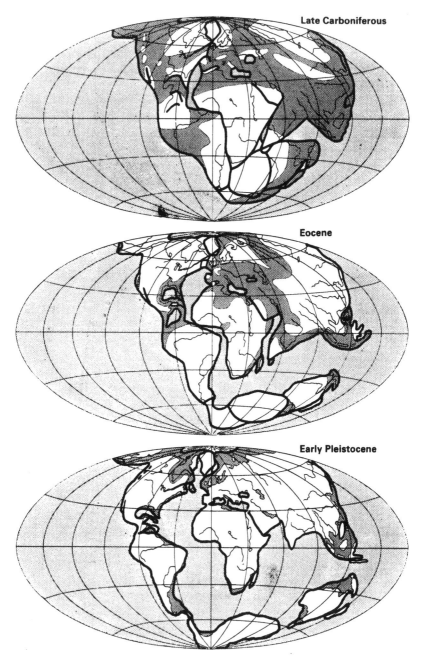

Figure 2.2 Wegener's reconstruction of the distribution of the continents during the periods indicated. Africa is placed in its present-day position to serve as a standard for reference. The more heavily shaded areas (mainly on the continents) represent shallow seas (after Wegener: from Holmes (1978).

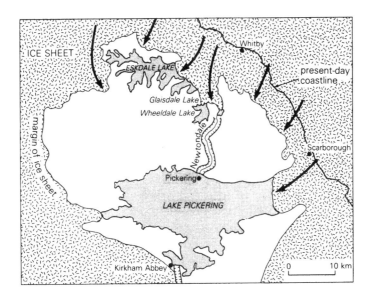

Figure 2.3 Glacial lakes in the North York Moors (after Kendall: from Sparks & West 1972).

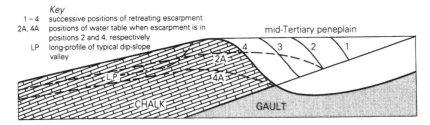

Figure 2.4 Fagg's hypothesis of dry valley formation (from Fagg 1923).

streams of the scarp slope, which benefited from the capturing, are also underfit. But the facts were twisted to fit the theory, instead of being used to produce a modified theory.

Conclusion

I want to put this debate into perspective. First, the cycle came at a very early stage in the development of geomorphology. Much that is known now was not known then. The surprising thing is not that the cycle can now be criticised, but that it endured for so long as a teaching tool.

Secondly, in view of this, perhaps the energy put into this debate is misplaced. Certainly the controversy has absorbed a great deal of introspective thinking in geomorphology, at least in Britain.

Thirdly, the cycle was never adopted on the scale that one might think. Davis's ideas left little impression in Germany, and in the Soviet Union and much of Eastern Europe geomorphology has proceeded without reference to the cycle. Even in Britain the cycle was slow to catch on. In 1924 Wooldridge observed that British workers had been slow in following the Davisian lead. In the United States, too, many geomorphologists lost interest in debating about peneplains during the 1930s.

Fourthly, even at the time that Davis was writing, much important work was being done in geomorphology and geology that was unrelated to the cycle. We have already noted Gilbert's (1914) work on fluvial processes. Other examples, just a few well known ones among many that could be quoted, are Wegener (1912) on continental drift (Fig. 2.2), Garwood on the glacial protection theory (1910), Kendall (1902) on glacial overflow channels (Fig. 2.3), de Geer (1912) on varves, Högbom on periglacial studies (1914), Fagg (1923) on chalk dry valleys (Fig. 2.4), Blackwelder (1933) on laboratory weathering experiments and Horton (1932) on drainage basin characteristics.

However, to see the full impact of the cycle, it is important to realise that the concept was applied in numerous other branches of geomorphology as well, so we turn to that next.

3 Other applications of the cycle concept

Introduction

As an idea in geomorphology, the cycle proved popular more or less immediately. During the next 50 years the concept was applied in eight geomorphological fields not specifically recognised in Davis's erosion cycle. As it happens, not one of these eight cycles has an important place in modern geomorphology, although King's theory remains a topic for intellectual debate in much the same way as Davis's theory does. Some of the criticisms of Davis's cycle listed in Chapter 2 can be levelled against these cycles; in several cases, conditions could not have remained stable long enough for the cycle to run its course; some are not true cycles, in that the end position is quite different from the initial one; and most geomorphologists today do not think in terms of cycles anyway. The eight are introduced here, not because of their current significance, but because they show that the basic idea of a cycle was so appealing that it stimulated further enquiry along the same lines for 50 years in many parts of the world and in many different geomorphological environments.

The eight cycles are described here in chronological order.

Cycles of erosion

The arid cycle

Aridity was Davis's second 'climatic accident', and he himself suggested a separate cycle for arid areas (Davis 1905). He based his ideas on the desert areas of southwestern United States, which led him to think that the cycle is initiated by block-faulting, a situation not true in all deserts. The arid cycle, although much criticised and not in use today, proved long-lasting and a detailed account of it appears in textbooks such as Holmes's *Principles of physical geology* (1978). The initial high plateau is subjected, in the stage of youth, to rare short-lived flash floods which, given long enough, cut gullies in it. These develop into wadis and canyons from which rock waste is spread over depressions. The uplands are bordered by steep escarpments which recede by wind erosion.

The stage of maturity may be said to begin when the mountains stand like islands, half submerged in their own waste products. This happens as the alluvial fans at the foot of the uplands grow and coalesce into bajadas which steadily increase in thickness as they encroach on the escarpments. Free faces of bare rock are prominent slope elements on the escarpments.

By the process of parallel slope retreat, the upland is gradually consumed during old age to leave a flat erosion surface as the end product. It is composed of bare rock, thinly strewn with debris, and smooth tracts of desert pavement and vast wastes of sand.

Little of this is now embodied in modern desert geomorphology. There are two points of interest, however. First, Davis himself conceded parallel retreat as a slope process, at least for rocky and stony semi-arid areas, a point not widely recognised. Secondly, the whole idea of an arid cycle is an important forerunner of a later cycle which attracted much more attention – King's pediplanation.

Mountain glaciation

Davis also regarded glaciation as a climatic accident, and in 1900 he himself suggested the possibility of glacial erosion producing its own cycle of events. This idea was taken up and elaborated by Hobbs (1910).

Hobbs's cycle is basically a scheme of stages in the destruction of an upland by the recession of cirques. In the first stage, the 'grooved upland' of youth still has much of the preglacial surface intact, supposedly exemplified by the Bighorn Range in Wyoming. The second stage – the 'early fretted upland' – is well displayed in the Snowdon massif in North Wales, where the smooth-topped ridge is being cut into by well proportioned cirques. As the cirques enlarge and recede, so a series of arêtes and horns is created, typical of the third stage, that of the 'mature fretted upland'. In Britain, this can be matched in the Cuillins of Skye. Hobbs then concludes the cycle with a stage of old age (his 'monumented upland') in which the breaking down of arêtes creates cols which allow ice to spill over from one cirque into another. Beneath the ice, the original floors may coalesce into a surface of glacial denudation, surmounted only by horns.

There was much support for the cycle. To quote but one example, Strøm (1949) thought that all Hobbs's stages could be found in the Jotunheim of Norway. However, it receives little attention from glacial geomorphologists today. Cotton (1942) summed up the position 40 years ago: 'Destruction of mountainous relief by glacial peneplanation is a theoretical abstraction in that no cases of such levelling down of mountains are known with certainty . . .' (ibid., p. 195).

Karst erosion

This is the only example of a cycle designed for just one rock type. Cvijic (1918) the originator of much karst terminology, and who was able to observe much magnificent limestone scenery in his native country, Yugoslavia, also proposed the idea of a karst cycle (Fig. 3.1). It is initiated by the uplift of a thick, pure mass of limestone underlain by

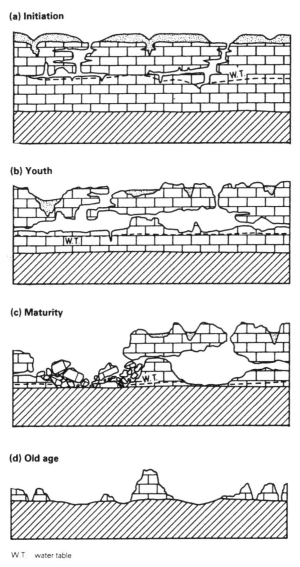

(a) Initiation

(b) Youth

(c) Maturity

(d) Old age

W.T. water table

Figure 3.1 The cycle of karst erosion (after Cvijic: from Small 1978).

an impermeable layer, and at first a well developed system of surface drainage will exist. However, during the stage of youth the development of sink holes by solution leads to the steady disruption of the surface system, and dry valleys and collapse features such as dolines become common.

The mature stage follows. A fully developed system of underground drainage, utilising fissures and caverns, comes into being, and the surface drainage disappears altogether. In late maturity some underground passages and caverns are revealed, and the subsurface streams approach the underlying impermeable layer.

During the stage of old age all subsurface caverns are revealed, and streams flow at the surface on the basal rock. Masses of limestone remain, but in time they, too, are removed, and the impermeable rock is left at the surface.

This is all very uncertain. Quite apart from the problems of whether all karst areas display this sequence of landforms and whether whole limestone areas have been denuded in this way, the sequence is not a true cycle. The removal of all the limestone in one cycle prevents the events being repeated merely by rejuvenation.

Marine erosion

Davis thought that marine action could have only limited effects by comparison with subaerial denudation, which can operate over the whole of the land surface instead of merely along its margins. Johnson rekindled interest in coastal processes and in particular in the idea, widely held in Britain in the 19th century, that wide surfaces of planation could be the product of wave erosion.

According to Johnson (1919) the youthful stage of the marine cycle would operate somewhat differently from one locality to another, depending on the detailed configuration and gradient of the initial coastline and on whether the cycle was begun as a result of submergence or emergence of the landmass. However, the general idea is that any irregularities of youth are smoothed out to give a fairly straight mature coastline, and during old age prolonged cliff recession exposes a wide wave-cut platform – the planation surface.

This is important, if only because the erosion surfaces so central to the work of denudation chronologists might have been formed by marine processes. This is discussed further in the next chapter, but for the moment we can note two points: first, the development of a very wide platform presumably requires a steadily rising sea-level; secondly, the complicated oscillations of sea level during the Pleistocene surely prevent any such cycle from running its course. Certainly, cyclic ideas play no part in the modern process approach to coastal geomorphology of, for example, Davies (1980).

Panplanation

Panplanation is different from all other cycles discussed in this chapter in that it is an alternative cycle under conditions of 'normal' subaerial erosion, or, at least, it is a significant revision of Davis's views as far as the later stages of the erosion cycle are concerned. Crickmay (1933) argued that the process of divide-wasting, involving the weathering away of interfluves and a progressive decline of slope angles, was not of primary importance in the formation of late-stage landforms. He envisaged lateral planation by meandering streams, rather than down-wasting, as the controlling old age process, and he thought that coalescence of floodplains would create an extensive erosion surface, the panplain.

The cycle of panplanation never commanded wide acceptance, for, although the process of lateral erosion by rivers is not in doubt, there are some problems. For example, the fact that meanders cut off limits the width that a floodplain can attain, unless the whole meander belt migrates laterally. Indeed, there must be some doubt as to whether vast panplains can be produced by coalescence of floodplains – the present-day features that can be attributed to lateral erosion, such as river terraces or broad floodplains that are veneered by alluvium, are not of any great extent.

Savanna erosion

Cotton's (1942) basic idea, shared by other geomorphologists, is that savanna landscapes, experiencing high annual temperatures and contrasting rainy and dry seasons, must have an erosion cycle different from the humid temperate areas. In Cotton's view, the steep-sided and often dome-like inselbergs are surrounded by true plains of erosion, the product of powerful lateral erosion rather as in the cycle of panplanation. The plains are not rock pediments (discussed in the next section), for these, according to Cotton, develop under conditions of greater aridity.

Whether or not Cotton's cycle would have been recognised today is difficult to say, for his ideas have been overtaken by three later events. First, King claimed that his cycle of pediplanation applied to savanna areas, a point also to be discussed in the next section. Secondly, in 1966 Pugh proposed an alternative erosion cycle for savanna areas which was based on the idea that uplift would lead to the removal of deeply rotted rock and which explained 'multi-cyclic' or 'dome-on-dome' inselbergs. Thirdly, the whole issue of cycles in savanna areas has been pushed to one side by the great interest shown in the origin of inselbergs, an interest which has proceeded without reference to cyclic ideas.

Pediplanation

Partly because Davis's arid cycle was inadequate in explaining the erosion surfaces of Africa (and, in fact, Arabia and Australia), King (1950) formulated a wholly new cycle for arid, semi-arid and savanna landscapes (Fig. 3.2).

King was a New Zealander who worked mainly in southern Africa. He identified two main elements in the African landscape which had to

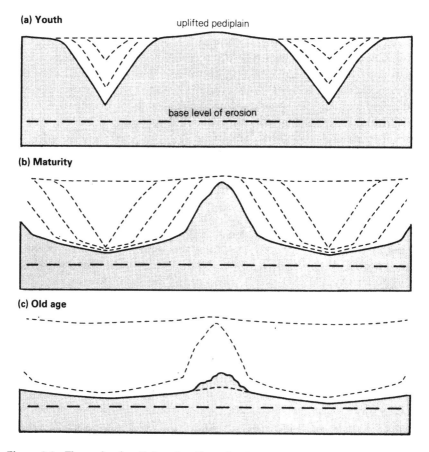

Figure 3.2 The cycle of pediplanation (from Small 1978).

be accounted for. The first was the pediment, an erosional slope cut into bedrock, with a gently concave profile characterised by low slope angles of up to about 7°. All the processes involved in its formation were grouped under the term 'pedimentation', although the concavity was suggestive of the dominance of running water. The second was the scarp, a steep slope of about 15–30° which bounded most upland blocks and which was separated from the pediment below by an abrupt break of slope. According to King, the scarps receded without change of angle, a situation he called 'scarp retreat' and which we commonly refer to as 'parallel retreat'.

King worked out a detailed cycle involving the usual stages of youth, maturity and old age, but in essence his ideas are as follows. An existing plain is uplifted, dissected, and the scarp slopes at the edges of hill masses retreat parallel to themselves. As they do so, the pediment is created and enlarged, and when all the hills have been consumed by backwearing, a low-relief landscape of coalescing pediments is left.

He went on to formulate in bold fashion a complete erosional history of southern Africa, based on this cycle. Several separate erosion cycles were identified, the earliest dating back to the Jurassic, and the erosion surfaces covered vast areas of southern Africa.

Working on a massive, subcontinental scale, he maintained that removal of material from the edges of the continent would lead to isostatic uplift and thus herald-in the next cycle. The emphasis on wearing *back* rather than wearing *down* meant that old, high-level erosion surfaces might be preserved and identified. He also attempted to correlate the effects of the cycles in Africa with those in South America, India, Australia and Antarctica.

However, King went much further than that. He saw his cycle as a genuine alternative to Davis's cycle of normal erosion, and as such he maintained that pediplanation was the basic process of landform evolution operating in all but glacial conditions. So here we have a cycle which is different from the other seven described in this chapter, in that it is supposedly of near-universal application and is in direct opposition to Davis's own near-universal cycle. It also involved long periods of geological time, just as Davis's did.

King's insistence that the cycle was appropriate to humid mid-latitude landscapes not unnaturally stimulated considerable interest in Britain. Even today attention is given to the identification of straight slopes, of pediment-like lowlands, and of the possibility that slope gradient remains constant as erosion proceeds. An example of such a line of thought appearing in a textbook within 10 years of King's publication is Dury's discussion (1959). For a landscape in the English Midlands he argues the case for identifying inselbergs, pediments and parallel retreat.

All in all, King's cycle of pediplanation has stimulated almost as much debate in Britain as Davis's cycle. The two are often discussed as competing all-purpose theories, just as King intended. However, since in modern geomorphology the whole idea of such grand all-embracing cycles has fallen into disfavour, King's cycle is regarded in much the same light as Davis's – something of an historical irrelevance.

But some of King's work is still of relevance today, even if his thoughts on cycles are not. There are two reasons for this. Firstly, his idea that slopes exhibit parallel retreat is a totally different interpretation of slope evolution from Davis's divide-wasting. To quote again from Dury (1959), the debate is encapsulated in the chapter title 'Wearing down or wearing back?'. If it is argued that even this is not a central issue today in slope geomorphology, then we can point to something else to show the present-day importance of his work. For, secondly, he produced perfectly plausible theories, notwithstanding that there are problems, for two highly controversial landforms which are still the subject of research – the inselberg and the pediment. His theories stand alongside more recent alternatives. Indeed, if one argues that there is a continuum of landforms that goes inselberg–mesa–bornhardt–koppie–tor (i.e. that these are basically all different versions of the same landform), then we can apply King's ideas to another much-debated feature, the tor.

Periglacial erosion

In 1950 Peltier proposed a theoretical cycle to account for periglacial landforms and especially those, such as altiplanation terraces, which have near-level surfaces (Fig. 3.3). The essential process is apparently cryoplanation – planation by frost. Peltier seemed to be aware that the relatively short duration of each mid-latitude periglacial period during the Pleistocene made it unlikely that a Davisian-type cycle could run its course, for he pointed out that the landscape would already be partly dissected and levelled. Indeed, the stage of youth begins with a pre-existing landscape, the slopes being converted by intense frost action into steep bare rock faces at an angle of 25–30° or more and undergoing a form of parallel retreat. Beneath these slopes, low-angle (15–20°) cryoplanation surfaces would develop, resembling steep pediments. Frost debris from the uplands could be solifluctued across these surfaces to accumulate in the valley bottoms. During the stage of maturity, the interfluves will gradually be consumed by the retreat of the rock faces. This will expose extensive cryoplanation surfaces, which will themselves be flattened out further by a combination of frost weathering and solifluction. By the stage of old age, all slopes will have declined to an angle of 5° or less, and their detrital cover will have been

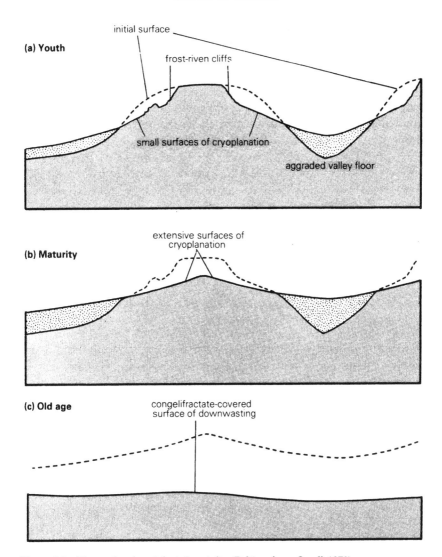

Figure 3.3 The cycle of periglaciation (after Peltier: from Small 1978).

comminuted to the point where wind transport can begin, sweeping the cryoplanation surfaces by the process of deflation and producing windswept pebble pavements.

Although the most recent of the cycles, Peltier's ideas have not survived. Periglacial environments are an important part of modern geomorphology, but cyclic ideas play little part in modern thought.

33

Conclusion

The concept of the cycle as proposed by Davis was so attractive that it was taken up by eight geomorphologists and applied to environments not envisaged by him. Not one of the eight cycles is of central importance in the subject today, however, and we can now go on immediately to look at the impact of the original normal cycle in Britain.

4 The response in Britain to the cycle: denudation chronology

Introduction

In this chapter we return to the mainstream of geomorphological thought, at least in Britain. Denudation chronology was the deliberate attempt to make use of erosion surfaces at different altitudes to reconstruct the geomorphological history of a region. By 'erosion surface' we mean an extensive flat or nearly flat area known to have been produced by erosion, and representing the end product of a cycle.

In Britain, the surfaces were nearly always interpreted as peneplains or marine platforms. The whole idea derives directly from Davis's cycle – the one led to the other. The erosion surfaces are, in effect, old age landforms; an attempt is being made to establish a sequence of events; the general approach is historical.

It is difficult to say when denudation chronology began. Planation surfaces were recognised in the 19th century, and, for example, what became known as the Mio–Pliocene surface was attributed to marine erosion. Davis himself maintained, in 1895, that it was a subaerial peneplain, only for Bury (1910) to revive the marine hypothesis. However, denudation chronology is not about the origin of just one surface, and probably the pioneer work on reconstructing a complete erosional history of a region was Baulig (1928) on the Massif Central in France.

It is also difficult to say when denudation chronology ended. In a sense it has not ended, for some geomorphologists still proclaim its virtues, and research is still being done which has a bearing on the debate. However, the stream of articles attempting to reconstruct denudation chronologies dried up in the mid-1960s, and its place in the centre of the stage was lost thereafter.

Even so, that is an impressive lifespan. It means that denudation chronology was the central theme, the end to which most research was directed – in short, the paradigm – for geomorphology in Britain for something like 30 years, having reached a peak of activity during the 1950s.

One effect of denudation chronology was to reaffirm the link between geomorphology and geology. The belief was that landforms were the best indicators of Earth history, and that geomorphology should provide the answers to questions about the most recent periods of geological time. The work reconstructed events that took place in the Tertiary and even the Mesozoic: further back than most modern studies in geomorphology. Also, the hiatus in deposition in the Tertiary in some parts of Britain and Europe encouraged geologists to believe that the gap could be filled on the basis of geomorphological evidence rather than the normal method of stratigraphy. In short, it was a deliberate attempt to use geomorphology as an aid to geology.

Denudation chronology had a subsidiary aim in addition to the identification, dating and interpretation of planation surfaces. It was to study the way in which the drainage system of an area has evolved. Through the identification of subsequent streams and the interpretation and dating of river captures, the original consequent pattern was reconstructed (Fig. 4.1) and the possibility of superimposition or antecedence investigated. The well known assumption of widespread

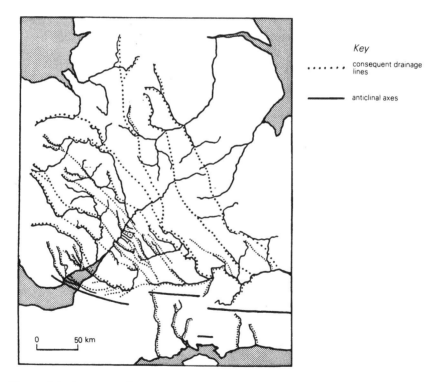

Figure 4.1 A suggested reconstruction of the former consequent drainage of the Midlands (after Ruckman: from Wooldridge & Morgan 1959).

superimposition from a formerly extensive covering of Chalk over southern England dates from this period.

The techniques used by denudation chronologists to identify erosion surfaces comprised a mixture of field observation and analysis of contour maps. Planation surfaces were recognised originally because they were obvious elements in some landscapes. However, they are prominent visually only in certain areas, such as Dartmoor and the Grands Causses in France. The other relevant field evidence was that, within a certain area, the summits of hills are often much the same height. These features, called accordant summit levels, were held to represent isolated remains of a formerly continuous surface at that height (Fig. 4.2). Some denudation chronologists backed up these observations with detailed surveys using instruments ranging from altimeters to levels. The most important mapping technique was the drawing of hypsometric curves for the areas under consideration. These are cumulative frequency graphs showing the percentage of the total area lying above, or below, certain heights. Where the graph flattens out, it means that an unusually extensive area occurs at that altitude, and that represents an erosion surface. This technique unearthed some erosion surfaces that were not obvious in the field, as well as confirming those that were.

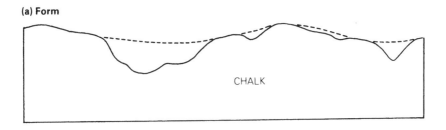

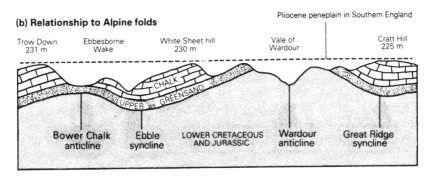

Figure 4.2 The Pliocene peneplain of southern England (from Small 1978).

The British scene

The classic work on denudation chronology in Britain was Wooldridge and Linton's *Structure, surface and drainage in South-East England* (1955). It was the culmination, and summary, of a body of research carried out for that part of Britain by Wooldridge and Linton themselves and by others, such as Sparks (1949) on the South Downs, Coleman (1952) in East Kent and Everard (1954) on the Hampshire Basin. The work is a complete book in itself and it examines many ideas and contains much local detail. The main conclusion, however, is the identification of four important erosion surfaces which are extensive over south-east England. In chronological order they are as follows.

(a)   A sub-Cenomanian (i.e. sub-Chalk) surface, much distorted by subsequent earth movements, and not in fact an important surface relief feature (Fig. 4.3).
(b)   A sub-Eocene surface, also much deformed by the Alpine orogeny, but an exposed and easily recognisable feature of parts of Chalk dip-slopes (Fig. 4.4).
(c)   The Mio–Pliocene surface, to explain the observation that the higher summits in the area show a marked accordance at about 700–900 ft a.s.l. (213–74 m). Wooldridge and Linton followed Davis in viewing it as a subaerial peneplain, not a marine surface.
(d)   The Calabrian marine surface. There is incontrovertible evidence of a marine transgression over much of south-east England in late Pliocene times – the Plio–Pleistocene or Calabrian transgression (Fig. 4.5), which produced widespread levelling at about 550–600 ft (168–83 m).

Although commonly regarded as the classic region for such studies, south-east England was not the only area to attract the attention of denudation chronologists. Surfaces were identified, and sequences established, for example, in Dartmoor, in Exmoor (Balchin 1952) and in the South Pennines (Clayton 1953). However, the main area outside south-east England was Wales.

In Wales, the researches of a number of workers since the last century resulted in a bewildering variety of erosion surfaces at different heights in different areas. The situation provoked a highly controversial discussion, but there was some measure of agreement on the authenticity of three surfaces:

(a)   A summit plain at about 600 m, often referred to as the Welsh tableland. Its origin attracted a number of theories, notably those of Jones (1951) and Brown (1957, 1960). In fact, the only certainty

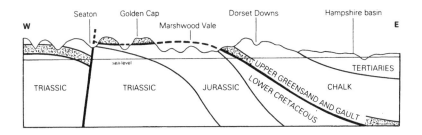

Figure 4.3 The sub-Chalk surface of southern England (from Small 1978).

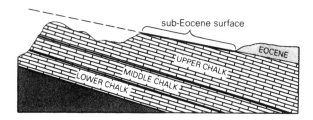

Figure 4.4 The sub-Eocene surface of southern England (from Small 1978).

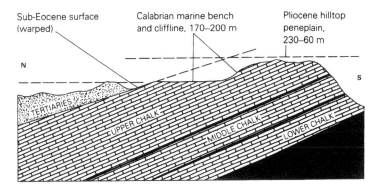

Figure 4.5 The relationship between the sub-Eocene surface, the Pliocene peneplain and the Calabrian bench in the central North Downs (from Small 1978).

is that it is post-Hercynian in age. By analogy with a similar morphological situation in Brittany, where the supporting sedimentary evidence is more conclusive, it is thought to belong to the Palaeogene, that is, to pre-date the Alpine orogeny.

(b) A Middle Surface at about 450 m.

(c) A Lower Surface at about 275 m.

Surfaces (b) and (c) in contrast, probably post-date the Alpine orogeny.

Not surprisingly, there have been many attempts to establish a comprehensive overall picture for Britain by comparing the sequences in different parts of the country. Sometimes the heights do seem to fit. For example, the altitude of the Lower Surface of Dartmoor (229–90 m) suggests that it correlates with the Mio–Pliocene surface in south-east England. But it is not as simple as that, for there are several reasons for believing that remnants of what was once a continuous surface might not be at the same height today. On more resistant rocks, the surface might be higher now than it is on softer rocks. The Alpine orogeny will, at the very least, have tilted and folded the pre-Alpine surfaces. In some parts of the country the surfaces will have been subjected to glacial erosion; in others they won't. Postglacial isostatic recovery will have tilted and deformed the surfaces. Parts of the classic area in south-east England belong to the area of long-continued downwarping around the southern part of the North Sea. All these issues complicate the business of correlation, and in particular they allow for the possibility that one former erosion surface now stands at different heights in different areas. An example of such a line of thought is Linton's (1964) attempt to correlate the established sequence in Wales with that in Scotland, such that the 600, 450 and 275 m surfaces in Wales correspond to the 950, 600 and 300 m surfaces, respectively, in Scotland. At this point, of course, the discussion is becoming highly speculative. However, as far as correlations are concerned it is only a beginning, for we have yet to consider results from abroad.

The world scene

The development of the concept of eustasy by Suess (1888) had a major influence on denudation chronology. Suess rightly concluded that worldwide synchronous changes of sea-level had occurred in response to the waxing and waning of ice sheets during the Pleistocene, and to movements of the ocean floors and the filling-in of ocean basins by sedimentation. The fact that the sea-level changes were worldwide and synchronous offered the chance for worldwide correlations of erosion surfaces on the basis of height alone.

However, in view of what has been said about the problems with building up a comprehensive picture for Britain, we need not delay unduly over imaginative attempts to correlate sequences in different parts of the world. But it is important to appreciate that denudation chronology was not a purely British obsession. Reference has been made already to work in the Massif Central and Brittany. Much is known also about sequences around the Mediterranean coast, and about the Ardennes, and Macar (1955) compared results from the Ardennes with those from the Appalachians in North America. Indeed, denudation chronology was almost as important in the United States as it was in Britain. The Appalachian Plateau became the classic area for study, and indeed the classic area for disagreement: the general consensus was something between three and five surfaces, but some workers claimed to recognise a dozen within a small area. This, however, did not deter Brown (1961) from attempting to correlate the Appalachians with Britain, much as Macar had done for the Ardennes. The denudation chronology approach lasted into the 1960s in the United States, too. For example, as late as 1962 Bretz identified two peneplains in the Ozarks.

Criticisms of denudation chronology

To most modern geomorphologists, denudation chronology is not relevant to the subject today. To some extent this is because the historical approach – the reconstruction of long sequences of events – is not in favour. The same point was made in connection with Davis's cycle. But that is not all. There are various specific ways in which denudation chronology is unsatisfactory, and these are listed below. It must be stressed that the approach is upheld by some academics, and they would dispute most, if not all, of what follows. The consensus of opinion, however, is that the following points constitute legitimate criticisms.

(a) Widely differing ages and modes of origin have been postulated for a single surface. The origin, dating and correlation of erosion surfaces is beset with difficulties and has produced a debate which is controversial and introverted even by geomorphological standards.

There are several reasons for this:

(i) The possibility that different remnants of an erosion surface will be at different heights in different places has been discussed already in the section on the British scene. Now it is significant that most authorities think that the Alpine orogeny reached its climax late in the Tertiary period – in the Miocene and even into the Pliocene. Few of the surfaces are thought to date from the Pleistocene, so the distorting of erosion surfaces by the orogeny just about rules out the possibility of straightforward correlation. Also the surfaces might well have been lowered by erosion and solution, making the heights almost meaningless.

(ii) Some surfaces, in certain places, might be structural. Of course, this cannot be the case when a nearly level surface is bevelled across dipping or folded rocks (Fig. 4.6), but it might be if the gradient of the surface is the same as the underlying structure (Fig. 4.7). This is how a structural surface is defined, but it might be a proper erosion surface anyway. Therein, in areas of horizontal or nearly horizontal rocks, lies the problem.

(iii) Some surfaces might be exhumed. In other words, the surface is cut, then covered and protected by later deposits, and then the deposits are removed to reveal the ancient surface. Add to this the possibility that the surface is later distorted, and some complicated reconstructions have to be done. This is the case with the sub-Cenomanian surface.

(iv) Any one surface could have various theoretical origins. It doesn't have to be a peneplain. It could be a marine surface, or a pediplain, or even conceivably the product of one of the other erosion cycles. The problem is partly that the surfaces, especially the older ones, are highly modified, but perhaps more important is the frequent absence of deposits on the surface. The whole question of interpreting superficial deposits is itself fraught with difficulties, but nevertheless it is true to say that the most successful reconstructions have been made where deposits such as marine gravels give supporting evidence. Such evidence is rare.

(v) There are several problems of dating, especially in areas where volcanic rocks are absent. In this situation, potassium–argon techniques cannot be used to establish some dates that can be used as reference marks.

(b) Assuming that the higher marine surfaces have been interpreted correctly, interesting problems of falls in sea level become apparent. For example, the height and dating of the Calabrian bench implies a fall of sea level of about 200 m since the beginning of the Pleistocene, and the cause is not related to the waxing and waning of Quaternary ice sheets. Now this is widely accepted as a genuine and interesting problem in geomorphology and is therefore not a criticism of denudation chronology, but the position is less satisfactory if higher, older marine surfaces are also authentic. For example, the need to explain a 600 m fall of sea level since the mid-Tertiary does pose an interesting question. As

Figure 4.6 A planed-off surface of steeply dipping Carboniferous Limestone at approximately 60 m O.D. near Paviland, Gower, South Wales (after Kay: from Small 1978).

Figure 4.7 The structural surface of Echinoid Chalk near Stockbridge, Hampshire. The line of hills in the background forms part of the 'secondary' (belemnite) escarpment of the Chalk (after Kay: from Small 1978).

it happens, there are some possible solutions, such as the concept of orogenic eustasy (Goudie 1977) whereby a local orogenic event could produce a worldwide fall in sea level.

(c)   The very existence of some of the surfaces, especially local fragments, is disputed. Often they are not conspicuous in the landscape. There is a reason for that, as explained in the earlier discussion on map techniques, but even these techniques merit comment. For many of the areas, the maps of the time had an instrumentally measured contour interval of 100 ft (30 m). There-fore, reliance cannot be placed on surfaces based on the intervening contours. Further, the actual values of the contours plotted influence the conclusions that can be drawn. For example, if there really was a flat surface between, say, 175 and 190 m (575–625 ft), it will not show up very well on contour maps because the 600 ft (183 m) contour line would go through the middle of it. On the other hand, a surface between 183 and 198 m (600–650 ft) would show up well, because map contours would define it at top and bottom.

(d)   Even if we take an optimistic view of denudation chronology and assume that all the main surfaces exist and all have been interpreted correctly, then we are still only concerning ourselves with a very small part of the total landscape – as little as 10% of the total area even in some of the classic regions. The remainder of the landscape – the vast majority, and including valleys and slopes – is ignored. This is not satisfactory.

(e)   Denudation chronology is more or less of no practical value; it has virtually no application.

Denudation chronology today

Denudation chronology is not a central element in modern geomorph-ology in Britain or elsewhere. In fact, it is difficult to assess its significance now. Some would defend it, while many younger students know nothing of it. What is true is that there have been few reappraisals of existing evidence, few detailed replies to criticisms or to alternative interpretations, and few new facts to add. It is not that all the results have been proved wrong, although some have and others are known to be highly speculative. It is more that the approach has fallen into disuse. With issues still unresolved, it has been replaced as geomorphology's paradigm.

However, old issues get resurrected from time to time. It was pointed out earlier that superficial deposits, where present, provide vital evidence as to the age and origin of a surface. Today in geomorphology in general, depositional features are studied more than formerly, and the sedimentologist plays a more important role than ever before. Therefore, research into deposits sometimes yields results relevant to problems in denudation chronology. Let us take one example. The Mio–Pliocene surface is often associated with a superficial deposit called clay-with-flints. This deposit should provide the decisive evidence, but unfortunately its own origin has attracted a vast and conflicting literature. It still does. Research continues on the clay-with-flints in its own right, quite apart from its relevance to denudation chronology, and on deposits at lower levels. This is so for the work of Hodgson, Catt and Weir (1967), and Hodgson, Rayner and Catt (1974), who discuss the clay-with-flints, show that no marine deposits exist at lower levels on the South Downs, and end with a scathing attack on denudation chronology.

To bring thought on denudation chronology right up to date, one must conclude with the work of Jones (1981). The book is a general modern geomorphology of south-east England, and as such deals extensively with denudation chronology. In drawing together new ideas and research, Jones expresses new views on the evolution of drainage and on the events of the Tertiary and Pleistocene. For example, his reinterpretation of the Tertiary stresses the following points.

(a) Tectonic movements began as early as the Cretaceous, and continued for most of the Tertiary, reaching a significant climax in the early Palaeogene (Palaeogene = early Tertiary = Palaeocene, Eocene and Oligocene). There are no indications of any truly stable intervals.

(b) Denudation was very rapid during the early Palaeogene.

(c) This tectonic instability resulted in a number of intersecting subaerial and marine erosion surfaces during the Palaeogene which formed a continuous but multi-faceted sub-Palaeogene surface. There is no evidence to support the concept of widespread planation during the Palaeogene.

(d) It is this complex sub-Palaeogene surface that, having been buried and then exhumed, now gives the Chalk back-slopes their general form. In other words, the Chalklands were largely fashioned in the Palaeogene.

(e) The sub-Palaeogene surface was later modified, or embellished, by late-Tertiary and Quaternary processes.

(f) It is highly unlikely that a widespread low-relief peneplain developed during the Neogene (= late-Tertiary = Miocene and Pliocene). There may well have been local development of erosion surfaces, but the old concept of the Mio–Pliocene peneplain needs to be abandoned.

Conclusion

Davis's erosion cycle is a good example of the historical approach to geomorphology since it concentrates on the sequence of events over a period of time. In Britain it led directly to the development of denudation chronology, which sought to reconstruct the erosional history of regions by using the evidence of erosion surfaces and drainage patterns. The link with the cycle is strong, for denudation chronology is also an example of the historical approach, and the erosion surfaces are assumed to be the products of a complete cycle of erosion. South-east England and Wales became the classic areas, although many other regions in Britain were also examined. As a general approach to geomorphology, denudation chronology dominated the subject until about 1960. It became popular also in other countries, notably France and the United States. Most modern geomorphologists would level a number of criticisms at denudation chronology, many of them also criticisms of the erosion cycle concept, and although the approach would be defended staunchly by many of the 'old guard' it has now fallen largely into disuse. Little work is now done on denudation chronology, but recent geomorphological research on the classic areas, such as that by Jones, would seem to indicate that a significant revision of interpretation is needed.

5 The Continental response to Davis's cycle

Introduction

Chapter 4 examined the British response to Davis's cycle of erosion. In this chapter we see how it was received elsewhere in Europe. There are several points of interest, because some of the European ideas were incorporated into British thinking, and even within Europe the response was not the same everywhere. First, however, it must be remembered that Davis's influence was not confined to Europe. In fact, we have already seen some examples of his worldwide impact. In Chapter 4 the popularity of denudation chronology in the United States was made clear. In Chapter 3 several examples of cycles were given, and these were proposed by a variety of workers in a variety of areas. Some of the names are worth stressing: as in all geomorphology, people and ideas go together.

In the United States, Fenneman (1936) had only minor reservations about Davis's teachings and, in believing that the study and interpretation of the records left by erosion constitute the larger part of the science of geomorphology, preserved the term 'normal erosion'. The main trend in the United States became the making of an inventory of the relief of the country within the framework of Davis's theories, and Fenneman compiled a physiography of the whole of the United States. In New Zealand, Davis's ideas came through strongly in Cotton's three volumes, one on normal geomorphology (1942a), one on climatic accidents (1942b) and one on volcanoes as landscape forms (1944). In southern Africa, as has been described, King (1950), although opposed to Davis on a number of counts, followed his lead on identifying planation surfaces, on reconstructing cycles, and on using the historical approach. In short, and in general, Davis's ideas were well received in English-speaking countries.

Adherents to the Davisian approach in Europe

The theme of most of what follows in this chapter is that Davis's basic ideas were opposed in most of Europe outside Britain. Indeed, geomorphology on the Continent proceeded along quite different lines, as we shall see. However, there was also a measure of support for Davis, and we shall look at that first.

This support was to be found in France. The French school of geomorphology maintained a rich vein of Davisian concepts. The influence began early. His theories were the basis for de Lapparent's widely read textbook (1907). The tradition was maintained in the work of Baulig, a faithful admirer of Davis. We have seen already that he pioneered denudation chronology in Europe (1928), a type of study he persevered with – for example, the investigation of rejuvenation features in north-west France in 1935, and further work on erosion surfaces in 1952. He also concerned himself (1938) with the correct use of the Davisian genetic terminology – consequent, obsequent, subsequent, insequent – and with slope development (1940). His work, not surprisingly, aroused interest in Britain, and his *Essais* (1950) were widely read. Another who preserved the Davisian line was Macar. His textbook (1946) was limited to 'normal' geomorphology. And finally we can point to de Martonne, in whose widely read textbook (1935) the ideas of normal relief and normal erosion are paramount. But at this point we reach the turn of the tide in France, for de Martonne had some reservations about Davis's outlook, and in fact he became interested in the influence of various types of relief as early as 1913. It was not until the 1940s (1940 and 1946) that he specifically took a stand in favour of zonal geomorphology. By the 1950s, in the work of, for example, Cholley, Dresch, Birot, Cailleux and Tricart, we can see in France a break from Davisian concepts and a deliberate move towards climatic geomorphology. Tricart, one of the main exponents of the climatic approach, speaks quite apologetically, almost sadly, about France's early involvement with Davis's doctrines. Climatic geomorphology arose out of criticisms of Davis's cycle (the normal cycle), and we will look in detail at it shortly. Before that, however, we must remember that Davis was subjected to criticism along quite different lines, during his own lifetime and while he was still active as researcher and writer.

Criticisms of Davis

Walter Penck

During the 1920s, while he was still publishing frequently, Davis attracted criticism and opposition from a German geologist, Walter Penck. Penck worked for many years in South America. He has been much misunderstood and misrepresented, and with good reason. He wrote in difficult prose that reads awkwardly even in the original German; the need for translation made his work obscure to most English-speaking geomorphologists; his ideas seemed weird and unfamiliar to the Davisian school; and he failed to publish a single comprehensive account of his views. Many of these problems have been overcome now. His main work (1924) has been translated into English (Czech & Boswell 1953), and good accounts of his ideas appear in several modern textbooks.

Davis and Penck saw geomorphology through different eyes. Davis was attempting the explanatory description of landscape. Penck, on the other hand, adopted the unusual standpoint that the main purpose of geomorphological research is to obtain information that might contribute to the understanding of crustal earth movements. He was particularly dissatisfied with Davis's assumption of very rapid uplift followed by a prolonged period of structural stability. Now, as we have seen, Davis appreciated that too, but he did not consider the implications in much detail. Penck, however, built his whole body of theory on the alternative assumption of uplift so protracted that the landscape would be eroded at the same time as it was being elevated.

Penck made a number of assumptions. First, he assumed that any slope, even if curved, is in effect made up of a number of straight slope segments, called slope units. Secondly, all slope units undergo parallel retreat. Thirdly, the rate of slope retreat is assumed to depend on gradient, steep slopes retreating quickly. These three assumptions were built into a theory of slope and landscape evolution that rested on one more vital idea. This was Penck's notion that the shapes of slopes, and thus the shape of the landscape in general, were determined primarily by the rate of river erosion, and that this in turn was determined primarily by the rate of uplift of the land. Three situations were visualised. First, a constant rate of uplift would produce a constant rate of downcutting by rivers, and the result would be straight valley-side slopes which together made up a landscape of *Mittelrelief* (medium relief). Slope steepness would depend on rate of incision. Secondly, an accelerating rate of uplift would give convex, or waxing, valley-side slopes and *Steilrelief* (strong relief). Thirdly, a decelerating rate of uplift would yield concave, or waning, slopes. When uplift ceased, as it would eventually if it was slowing all the time, a flat or gently undulating surface called an *Endrumpf* (terminal surface) would remain.

Penck's next important idea was that of slope replacement from below. As a given straight slope unit retreats parallel to itself, it is replaced from below by another straight slope of gentler gradient (Fig. 5.1a). Since the steeper slope will retreat more quickly than the gentler one, the gentle slope will eventually replace the steeper one above it.

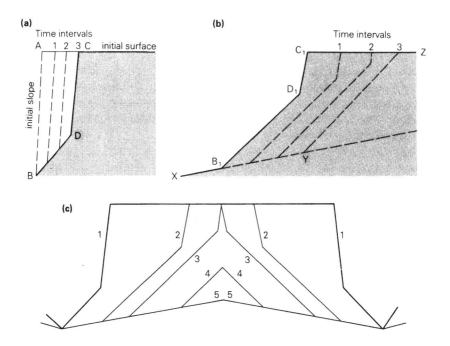

Figure 5.1 Penck's idea of how slopes might develop in homogeneous material. Unimpeded removal, but no erosion, is assumed at the base of the slope (from McCullagh 1978).

Meanwhile, the gentle slope itself will be being consumed from below by an even gentler slope (Fig. 5.1b). The result is a long-term tendency towards slope decline, albeit for reasons quite different from those envisaged by Davis.

The final item of interest in Penck's work is his concern with flat or gently undulating surfaces. In this context his main criticism of Davis was that in the erosion cycle the peneplain was always the end product. Penck had his own theory, that of the *Primarrumpf* (initial surface, or primary peneplain). He postulated the situation where a flat or gently undulating surface is uplifted and then eroded so that the flat relief remains: lowering is parallel to the original plain. As he points out, the landform looks like an end peneplain, but in fact it is at the beginning of a sequence of landforms. Penck's final theory concerns the formation of *Piedmonttreppen*. The full analysis is complex, but the gist of the idea is that the effect of intermittent uplift on a *Piedmontflache* (mountain-foot flat) will be the creation of a 'piedmont staircase' – a series of extensive flats at various levels called *Piedmonttreppen*.

At the time of their publication, Penck's views did not command wide acceptance, and for a long time afterwards were attacked strongly by leading American and British geomorphologists. They included Davis, whose defence of his position included visiting Germany, writing some articles in German, and issuing his own critique of Penck's work (1932). The debate continued for many years, and in 1940, for example, the divergence of views between Davis and Penck on the evolution of slopes was conceived as a subject for a special discussion by American geographers.

Criticisms of Penck's ideas appear in a number of textbooks, and it is not proposed to go over well trodden ground again here. Probably the one point that must be stressed is that one must dispute the idea that slope form and gradient are determined primarily by rates of river erosion. Other factors, which Penck regarded as subsidiary, are important: rock type, structure, vegetation, weathering and transportational processes.

But times change, and since 1945 it is Davis rather than Penck who has been under fire. Indeed, some of Penck's ideas, such as slope replacement and parallel slope retreat, are now seen as important. It is also worth noting that there are some similarities between the work of Davis and Penck. Both produced models of slope and landscape evolution, although Penck's made no use of the cycle concept; both used the deductive approach, making assumptions and then inferring logically how landforms would change with time; Tricart (1974, p. 21) sees Penck as following the Davisian line in his theory of the *Piedmonttreppen*, or Piedmont staircase; and Penck's *Endrumpf* surely smacks of Davis's peneplain.

In retrospect, the debate engendered by the contrasting views of Davis and Penck seems to be of more historical than vital contemporary geomorphological significance. To some extent their differences of view can be put down to the fact that they were the products of different environments. Penck came from recent fold mountains in Europe and worked in the Andes, while Davis hailed from the heavily denuded old mountain stubs of New England. It is probably true that no single theory of hillslope evolution is applicable to all environments. The search for such a theory, and one for landscape evolution in general, has been called off.

Climatic geomorphology

In Europe outside Britain and France – in Sweden, Germany, Poland and the Soviet Union, for example – geomorphology progressed more or less without reference to Davis's erosion cycle. Their criticism, reduced to its simplest, was that different climates produce different processes which in turn produce different landforms, a fact not stressed by Davis in his unifying cycle. In its place they adopted an alternative theoretical approach which we now call climatic geomorphology. Whole books have been written on this one subject, for example, Tricart and Cailleux (1972), but the theme of the approach is that distinctive climates have distinctive assemblages of processes which result in different assemblages of landforms. Some of the examples seem rather obvious: glacial landforms in the glacial zone, periglacial landforms in the periglacial zone, and so on. Every phenomenon or process whose global extension is more or less conformable to latitude is termed 'zonal'. But the analysis is not as simple as that, because processes occurring in several or even all zones (the azonal phenomena, e.g. river or wave action), do not necessarily produce the same landforms everywhere. This is because the intensity and frequency of occurrence of the process may vary from one part of the world to another, so that distinctive assemblages of landforms are still produced. The dominant influence of climate on landform is then underlined by considering the indirect impact of climate, via soils, vegetation and animals. This emphasis on biological processes is interesting because it is a traditional weakness of geomorphology in Britain. The end product of the climatic approach is the identification of a number of so-called morphoclimatic zones (Fig. 5.2) of the Earth, each with its distinctive climate, processes and landforms.

It is in Germany that we find the original development of climatic geomorphology. For example, at the very moment that Davis was elaborating his theory, von Richthofen (1886) was giving equal importance to the climatic factors and the geological factors in the development of relief. This early tradition was not lost, and Davis's success remained minor in Germany. Climatic geomorphology was continued in the work of Machatschek, whose early writings were contemporary with the mature publications of Davis and Penck, and who cautiously accepted the significance of climate in determining landform in the 1940s; and in the work of Troll, published mainly in the 1940s and 1950s. By about 1950 the approach was also gaining strength in France, a trend underpinned in the years that followed by the work and writings of Birot, Tricart and Cailleux.

As one might expect, climatic geomorphology has generated its own body of debate. For example, there is a fundamental disagreement between Tricart and Cailleu (1972) and Derbyshire (1973) as to whether climatic geomorphology is in opposition to Davis's cycle or a product of it. The line of thought presented so far in this chapter derives from Tricart (Fig. 5.3), who views climatic geomorphology as a criticism of Davis's work. In complete contrast, Derbyshire sees Davis's treatment of the landforms of the humid mid-latitudes as the first attempt to relate a specific suite of landforms to a specific type of climate and thus as the origin of climatic geomorphology. Looked at in this way, the

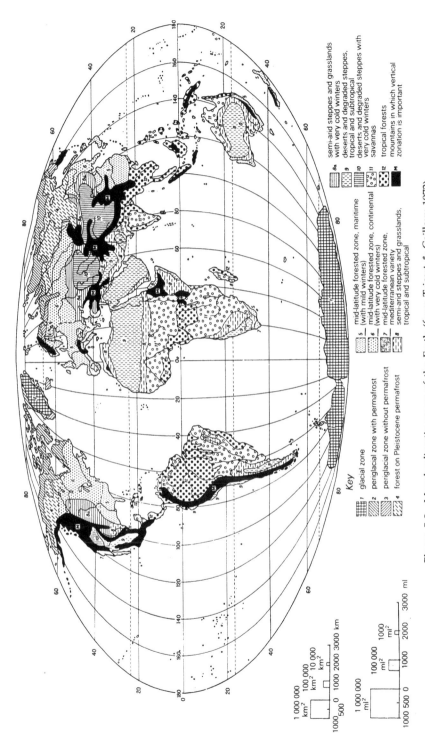

Figure 5.2 Morphoclimatic zones of the Earth (from Tricart & Cailleux 1972).

Key

1 glacial zone
2 periglacial zone with permafrost
3 periglacial zone without permafrost
4 forest on Pleistocene permafrost

5 mid-latitude forested zone, maritime (with mild winters)
6 mid-latitude forested zone, continental (with very cold winters)
7 mid-latitude forested zone, mediterranean variety
8 semi-arid steppes and grasslands, tropical and subtropical

8a semi-arid steppes and grasslands with very cold winters
9 deserts and degraded steppes, tropical and subtropical
10 deserts and degraded steppes with very cold winters
11 savannas
12 tropical forests
13 mountains in which vertical zonation is important

Figure 5.3   J. Tricart, the French geomorphologist (from Brunsden & Doornkamp 1977).

applications of the cycle concept to other environments (Ch. 3) represent a continuation of climatic geomorphology, although it must be stressed that not all the early contributions to the climatic approach were based on the cycle.

Over the years, there have also been many criticims of climatic geomorphology. The literature is vast, but in summary one can say that the critics have stressed the following points:

(a) As discussed by Stoddart (1969), it seems to be doubtful whether fluvially controlled landscapes formed in different climates are as distinctive as is sometimes maintained.

(b) Rock type can dominate climate. For example, in Britain, glaciated Chalk looks much the same as unglaciated Chalk. Also, karst geomorphologists, long preoccupied with explaining the apparently distinctive assemblages of landforms developed in tropical and temperate areas, now stress the differences between the limestones rather more than they used to, and the differences between the climates rather less.

(c) The reality of environmental change during the Pleistocene over most of the Earth's surface makes it rather unlikely that present-day processes will necessarily have formed present-day landforms. This topic will be discussed in more detail in the next chapter. It casts doubt on the identification of morphoclimatic regions, however, where the association of form and process is an implicit assumption.

(d) Crude climatic statistics and parameters were used. This rather highlighted the added criticism that landform is acted on by process, not by climate. Climate alone does not determine process. Therefore, climate is one step removed from reality.

(e) Equifinality is always a problem – in different areas, different

54

processes can produce similar landforms. Indeed, some of the 'regions' may be rather similar. In terms of general surface form, for example, arid and periglacial environments have several similarities.

(f) There was, and indeed still is, a lack of information in equal detail over the Earth's surface.

(g) It transpires that there are few truly diagnostic landforms – in other words, surprisingly few landforms are restricted to just one morphoclimatic region. Inselbergs, for example, supposedly diagnostic of deserts, in fact occur also in semi-arid and savanna environments and possibly elsewhere.

(h) The scale is gross, and the practical value of identifying vast 'regions' must be questioned.

(i) It is doubtful whether the regions are sufficiently homogeneous to be distinguished adequately.

(j) Changes wrought by man are not taken into account.

An interesting refinement of climatic geomorphology is the climato-genetic approach of Büdel (1963, 1969). In his opinion, the two main obstacles to the successful identification of morphoclimatic regions were climatic change and changes caused by man. Büdel distinguished three generations of study: dynamic geomorphology concerns the study of particular processes; climatic geomorphology considers the total complex of currently active processes in their climatic framework; and climato-genetic geomorphology involves the analysis of the entire relief including features adjusted to the present climate and features produced by former climates. The approach therefore concedes that the present is to some extent governed by the past. Büdel recognised seven climato-genetic zones (Fig. 5.4). In each zone a distinctive pattern of processes should operate at the present time reflecting the macro-climate, but variations within each zone can arise from the effects of factors other than climate, notably rock type, relief, soil and vegetation cover, and man.

In spite of the ten reservations listed above, climatic geomorphology is still important in Continental Europe, and it is now accepted as an important contribution to geomorphology in Britain, too. As evidence of its development in this country we can point to the collection of essays edited by Derbyshire (1973) and the translation into English of Tricart and Cailleux's textbook (1972) and Büdel's great work (1982). Therefore, as a criticism of Davis's cycle and as a viable alternative theoretical framework, we must regard climatic geomorphology as important, comprehensive and possessed of long life.

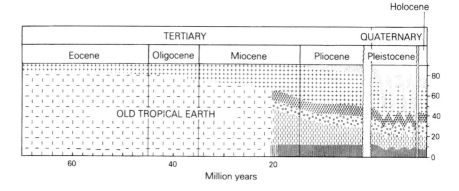

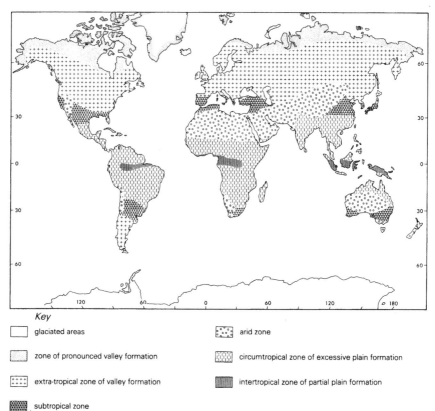

Key

☐ glaciated areas

▦ zone of pronounced valley formation

▦ extra-tropical zone of valley formation

▓ subtropical zone

▦ arid zone

▦ circumtropical zone of excessive plain formation

▥ intertropical zone of partial plain formation

Figure 5.4 Climato-genetic zones, based upon Büdel (1963, 1969). The upper diagram indicates the way in which the pattern of zones may have changed over time. Three horizontal scales are used to represent the durations of the Holocene, Pleistocene and Tertiary respectively (from Gregory & Walling 1973).

Structural geomorphology

As explained earlier, it is generally reckoned by geomorphologists that Davis, having identified structure, process and stage as the three main controls on landform development, failed to do justice himself to structure. Structural geomorphology has developed, therefore, as a deliberate attempt to make up lost ground, to redress the balance. Tricart, however, sees it as a complement to climatic geomorphology (1974, pp. 1–5).

In fact, structural geomorphology has not developed as a particularly distinctive viewpoint. Perhaps this is because it consists of several loosely allied threads. First, there is the simple matter of recognising that rock type and structure are important influences on relief. This is what Davis neglected and is the gap filled by, for example, Sparks's *Rocks and relief* (1971). Secondly, many landforms are structural in that they have been formed principally by structural movements in the Earth's crust: this is basically the subject matter of books such as Tricart's *Structural geomorphology* (1974) and Twidale's *Structural land-forms* (1971). It is this topic which has received such a boost from the rise of plate tectonics. Thirdly, the concern with rock type and structure has led to the study of the relief developed on certain rock types, notably limestone and granite. Fourthly, one can point to the importance of tectonic activity, traditionally a strength in Hungarian geomorphology and the subject of at least one modern British textbook – Weyman's *Tectonic processes* (1981). And finally, a modern view is that a vital key to understanding causality in geomorphology lies at the interface between rock and process, emphasising rock strength and the properties of materials, for example, Yatsu (1966) and Whalley (1976). But now we are beginning to overlap into something to be discussed in more detail in Chapter 7. Let us conclude by saying that a legitimate criticism of Davis is that he appreciated but did not emphasise the role of structure in explaining landforms, and that the omission has now been more than rectified.

Conclusion

Davis's erosion cycle was generally well received in Britain and other English-speaking countries, and also in France until about 1950. His ideas made little impact in Europe in general, however, and in Germany, Eastern Europe and the Soviet Union, for example, geomorphology has developed more or less without reference to the cycle. During his own lifetime, one of Davis's main critics was Penck, a German geomorphologist who produced his own theory on the evolution of landforms. Penck's work contains some unusual ideas but also many that have been incorporated into the current consensus of opinion. Partly as a result of criticisms of Davis, European geomorphology has yielded two approaches to the subject that are undoubtedly important today – climatic geomorphology and structural geomorphology.

6 Pleistocene geomorphology: the impact of environmental change

Introduction

The main theme of the previous four chapters has been that the historical approach (i.e. the investigation of sequences of events and the evolution of landscape) has been an enduring paradigm in geomorphology but that now it has fallen into disfavour. There is one exception to that trend – Pleistocene geomorphology is extremely important today, and arguably its main aim is to establish the sequence and dating of events during the Ice Age. In a sense, therefore, it is the last survivor of the historical approach. That is one reason for introducing the topic at this early stage in the book. The other reason is that the explanation of many landforms relies on a knowledge of it. Many authors consign their chapter on the impact of climatic change to the end of the textbook. I don't agree with that: it should come at the beginning. One cannot understand the rest of geomorphology without it.

This chapter is not going to be the usual bank of information about glacial and periglacial landforms and processes. Nor is it a comprehensive account of the chronology of the Pleistocene. Instead, it will examine some of the implications of recent findings for geomorphology. The emphasis will be on ideas and concepts, not facts. For the facts the reader is directed to the numerous accounts in standard textbooks, and for specialist information to the many excellent advanced textbooks such as Embleton (1972), Embleton and King (1975), Goudie (1977), Paterson (1969), Price (1972), Price and Sugden (1972), Sparks and West (1972), Sugden and John (1976), Washburn (1973), and West (1977).

Pleistocene chronology

The term 'Pleistocene' refers to a geological system, corresponding to what the layman calls the Ice Age. The period since the end of the

Pleistocene is called the Recent, and together they make up the last geological era, the Quaternary. The start of the Pleistocene is difficult to date exactly, not least because the onset of cold conditions does not seem to have happened at the same time all over the world, but in Britain we can deal with a convenient round number and say that the Ice Age began about 2 million years ago. To judge from discoveries made in East Africa and elsewhere, this is also about the time that man appears on the Earth as a recognisable animal, so the Ice Age is also the Age of Man. This is clearly recognised by the Russians in their use of the term 'Anthropogene' for the Pleistocene. Man's appearance at the start of the Ice Age may well not be a coincidence, since elsewhere in geological history there is evidence that sudden environmental change is synchronous with equally sudden evolutionary advances.

The Ice Age did not consist of a single refrigeration of climate. Instead, climate alternated between cold periods, which we normally call glacials, and warmer, temperate periods called interglacials. The exact number of glacial periods is in doubt. Traditional teachings, based on the classic Penck and Brückner model (1909) derived from work in the Alps, spoke of four – Günz, Mindel, Riss and Würm – but this is almost certainly an underestimate, and some modern sources refer to as many as 60 cold periods. The problem is partly one of definition. Climate was constantly getting colder and then warmer again by widely varying amounts both during and between glacials. Recent evidence from deep-sea cores shows that the degree and frequency of climatic change was very great at times during the Pleistocene. Indeed Bowen (1978) has likened the new ideas to plate tectonics in terms of how revolutionary they are. Therefore, we have to decide how cold it must get to be called 'glacial', and we have to decide whether two or more closely spaced cold periods represent several separate glacials or just one. It is difficult to evaluate all the possibilities. The majority view today is that there is strong evidence for six cold periods in East Anglia, and there may well have been more.

The terms Günz, Mindel, Riss and Würm are not used today for the glaciations of Britain. There is a reason for that. The classification was developed in the early 20th century for events in the Alps, and it is now recognised that it is by no means certain that cold periods in the Alps corresponded in either timing or number with those in Britain. The same argument applies to attempts to correlate the British sequence with that in North America, or anywhere else. So Britain has its own sequence and its own terminology. Actually, it is not even certain that cold periods were synchronous within Britain. The longest and most continuous sedimentary record is found in East Anglia, probably because it is an area of long-continued downsinking, and the British sequence is based on the East Anglian evidence.

The first three of the six cold phases did not result in glaciation as far as is known, but the fourth and fifth – the Anglian and the Wolstonian – produced extensive glaciations. They were comparable in size, although in general it is the Anglian which represents the maximum southwards extent of ice in Britain (Fig. 6.1). This is the 'Thames–Severn' line so often referred to in O-level textbooks, although it could be that the ice extended further south. The familiar event in the Wolstonian glacial was the formation of Lake Harrison in the English

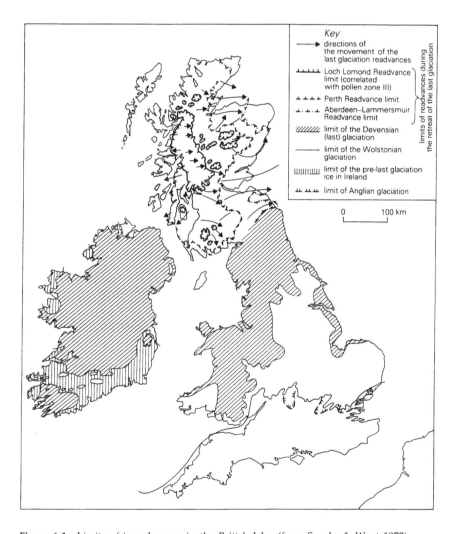

Figure 6.1 Limits of ice advances in the British Isles (from Sparks & West 1972).

61

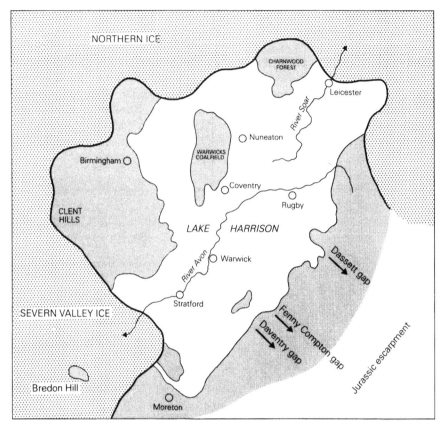

Figure 6.2 Lake Harrison in relation to present drainage (from Sparks 1972).

Midlands (Fig. 6.2). The sixth and last glacial, the Devensian, was not so extensive, although it did reach well to the south in Wales (Fig. 6.1).

The intervening interglacials also have names. The last interglacial, between the Wolstonian and the Devensian, is called the Ipswichian, the one before is the Hoxnian, and the one before that is the Cromerian. At their warmest, all were slightly warmer than present-day conditions – summers in the Ipswichian were about 2–3°C warmer than today – but in general we can picture conditions during the interglacials as being similar to today's. Indeed, in view of the likelihood of another cold period sometime in the future, it might be more accurate to refer to the present day as an interglacial rather than the postglacial period.

Much is known about the dating of some of the glacials and interglacials in Britain, although there is some measure of debate about nearly all of the dates. In general, the most recent periods can be dated with most certainty. Taking the general consensus of opinion, and working in convenient round figures, the following picture emerges:

350 000 BP (= Before Present)	Mid-Cromerian interglacial
250 000 BP	Mid-Anglian glacial
150 000 BP	Mid-Hoxnian interglacial
125 000 BP	Mid-Wolstonian glacial
100 000 BP	Start of Ipswichian interglacial
75 000 BP	Ipswichian interglacial ends
	Last (Devensian) glacial begins
25 000 BP	Devensian glacial at its coldest
13 000 BP	Rapid deglaciation of the entire country
12 000 BP	End of late-Devensian Zone I, a tundra phase
10 750 BP	End of Zone II, the Allerød Interstadial, a temporary warm phase
10 250 BP	End of Zone III, the Highland Readvance, a period when a small ice sheet re-formed in Scotland, cirque glaciers temporarily returned to Wales and the Lake District, and southern Britain reverted to tundra conditions (Fig. 6.1)

A mere 10 250 years ago, therefore, permanent ice sheets and glaciers last disappeared from Britain, and this is taken as the 'official' end of the Devensian and the beginning of the postglacial period, or Flandrian. Several points are worth noting about the figures. First, little or nothing is known about what happened during the first 1.5 million years of the Ice Age in Britain. Secondly, the entire sequence is geologically extremely recent and lies well within the period when prehistoric man was inhabiting Britain. Thirdly, Britain was actually under ice for only quite a small part of what we call the Ice Age. Fourthly, deglaciation was rapid at the end of the Devensian, and this is important geomophologically since nearly all the eskers, kames and other fluvio-glacial landforms now visible in Britain date from these retreat stages.

The postglacial period is short, and, taking a long view, highly atypical. It is classified by continuing the Zones begun for the late-Devensian. The present day is Zone VIII. Some of them are geomorphically important. For example, the Atlantic Period (Zone VIIa, 7000–5000 BP) is known to have been mild and wet compared to the present day. Therefore, if Dury's theory about misfit streams and dry valleys being due to a former period of high rainfall is correct this could be the period concerned. The details of climatic change during the last 2000 years are known with considerable accuracy, and again some of them are of interest geomorphologically. For example, the period 1550–1850 was unusually cold and is often called the 'Little Ice Age'. The glaciers of the mountains of Europe reached their most advanced positions for 10 000 years and left well marked terminal moraines.

Fossil landforms

One effect of environmental change during and since the Pleistocene is that landforms exist today that were formed by processes that no longer operate there. Present-day processes are slowly modifying the landscapes, but the landforms are easily recognised and are essentially unchanged. This is a well established idea in geomorphology, and one can read of fossil landforms, of relict landscapes, and of present-day landforms being out of phase with present-day processes.

An obvious example in Britain is upland glaciation. The glacial origin of cirques, arêtes, etc. is not in question, the landscape is being only very slowly modified by present-day processes, so the mountain scenery of Snowdonia, the Lake District and so on represents fossil glacial landscapes. Another example, not so obvious but just as certain and just as important, is that southern Britain is essentially a relict periglacial landscape. Putting to one side for the moment the debate as to how far ice did get in Britain, the area south of the Anglian and Wolstonian limits must have been subjected to repeated phases of periglacial activity, and even more extensive areas were affected during the Devensian. So one can look with confidence for landforms that are – or, in some cases, might be – of periglacial origin in southern Britain, and sure enough they are there in large numbers: tors, dry valleys, asymmetrical valleys, rock streams (Fig. 6.3), patterned ground (Fig. 6.4), involutions, ice wedges, solifluction deposits, convexo–concave slope profiles, pingos, cambering, gulling, valley bulging and possibly others. These landforms are not being formed today, so the conclusion seems inescapable – southern Britain is a fossil periglacial landscape. This is one reason why periglacial geomorphology is now seen as so important in Britain.

Relict landscapes are not confined to Britain, nor even to the mid-latitudes. For example, there is strong evidence that much of the Sahara Desert is a fossil landscape formed under pluvial (i.e. rainy) conditions during the Pleistocene. The subject is discussed in advanced textbooks on hot deserts such as that by Cooke and Warren (1973) and by Goudie and Wilkinson (1977), and there is also a good section in Hilton's well-known school textbook (1979, pp. 169–74). The essence of the argument, however, is that there are landforms and other features in the Sahara which are almost certainly not being formed under present-day conditions and which probably date from former pluvial periods. These include wadis, whole drainage systems, river terraces and inland deltas and lakes (Fig. 6.5). Some of the evidence dates the last pluvial period to 35 000–15 000 BP or thereabouts, in which case it is synchronous with the last mid-latitude glacial. It suggests a southward shifting of all the climatic belts, so that the Sahara acquired

Figure 6.3 An accumulation of sarsen stones in Clatford Bottom, Marlborough Downs, Wiltshire. The stones rest on a considerable thickness of coombe rock, and have been transported down slope to their present positions by solifluction in the Quaternary (from Small 1978).

Mediterranean characteristics. But this is not certain, and the climate may have been more of a savanna type, which would suggest a northwards shifting of climatic belts under warm interglacial conditions.

There is one tropical environment which seems not to have been affected by significant climatic change during and since the Pleistocene, and that is the humid tropics, lying astride the equator. The reason for that is an interesting climatological discussion, but for the geomorphologist the significance is that since climate has remained more or less unchanged for at least 2 million years, present-day landforms *will* be in phase with present-day processes. As has been seen, this happens only too rarely in other regions, and helps to explain the recent interest in equatorial landforms summarised by Thomas (1974, 1977).

One final point about process and form being out of phase. Recalling the examples where this is the case, the discussion was based mainly on erosional landforms. However, the same can apply to depositional landforms. For example, in Britain glacial and periglacial processes in

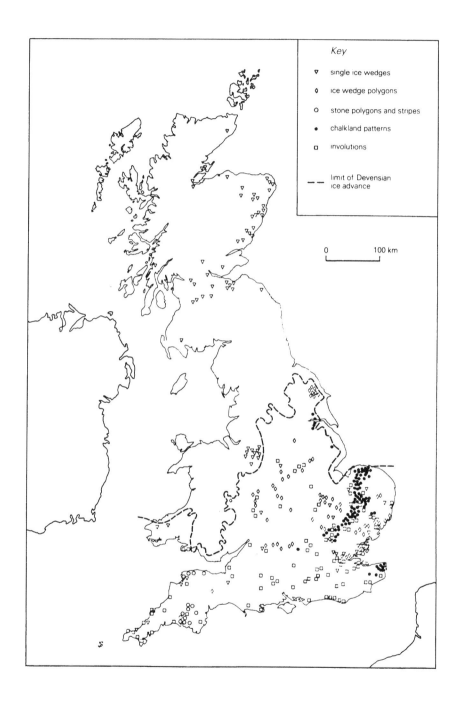

Figure 6.4 The distribution of periglacial patterned ground in Britain (from Sparks & West 1972).

the Pleistocene laid down many depositional landforms, and these are now out of phase with present-day processes. Over large parts of Britain today, streams and the sea are acting as agents of transport over fossil depositional landscapes. This is one cause of instability in the landscape: the landforms are being acted on by processes other than the ones which formed them.

Some implications of the discrepancy between form and process

So far it has been established that, at least in the mid-latitudes and the sub-tropics, it is common to find process and form out of phase. Now quite apart from being of interest in its own right, that has some important implications. We have met two already. First, in areas currently being affected by Davis's 'normal' processes, these processes have not been operating very long – probably about 10 000 years at most. Secondly, climatic geomorphology relies implicitly on an association between present-day form and present-day process, so the approach has a weakness on that score. It is now proposed to discuss some other implications.

We are forced to change our ideas as to how some landforms are formed and how some processes operate. Let us follow a line of thought on a familiar topic. How does ice erode bedrock? The answer is important because, in upland Britain and elsewhere, vast landforms over vast areas are involved. And there is a problem straight away, because most of the rocks forming the mountains are harder than ice. The orthodox answer is that two processes are involved, abrasion and plucking, and even though it is stressed that abrasion involves the action of rock on rock and that plucking is most effective when the rocks are jointed, one is left with the uneasy feeling that there must be more to it. And there is. One point not directly related to climatic change is that temperate glaciers often have meltwater streams at their bases, and these streams act as lubricants so that the ice can move over and abrade the bedrock, and are also capable of erosion themselves. But we have not used our knowledge of climatic change yet. Textbooks usually give the impression that cirques, say, are formed by a glacier developing in the incipient cirque, eroding it, and then going away. But we know this is not so, for we know that cirques in Britain were occupied by ice on at least three separate occasions and quite possibly more. So whatever cirque-glaciers can achieve by way of erosion, they did so at least three times. Multiple glaciation means multiple opportunities for erosion. Next, what happened between one glacial and the next? The answer is, a periglacial period, followed by an

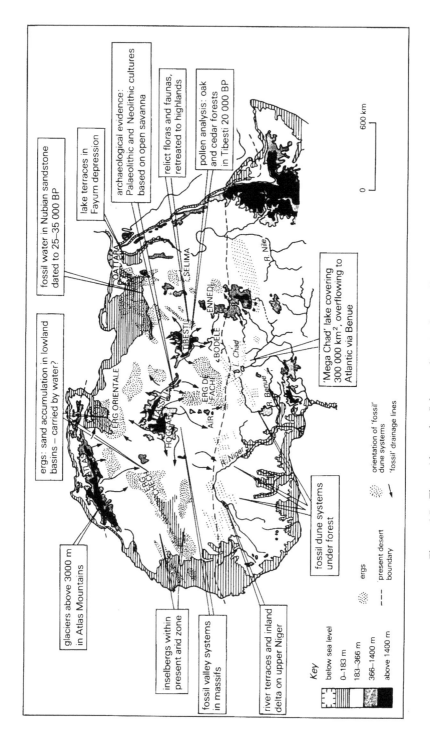

Figure 6.5 The evidence for Saharan climatic change (from Hilton 1979).

fossil water in Nubian sandstone dated to 25–35 000 BP

lake terraces in Fayum depression

archaeological evidence: Palaeolithic and Neolithic cultures based on open savanna

relict floras and faunas, retreated to highlands

pollen analysis: oak and cedar forests in Tibesti 20 000 BP

ergs: sand accumulation in lowland basins – carried by water?

'Mega Chad' lake covering 300 000 km² overflowing to Atlantic via Benue

glaciers above 3000 m in Atlas Mountains

inselbergs within present arid zone

fossil valley systems in massifs

river terraces and inland delta on upper Niger

fossil dune systems under forest

QATTARA

SELIMA

TIBESTI

ENNEDI

BODELE

ERG ORIENTALE

ERG DE FACHE

HOGGAR

AIR

ERG CHECH

ATLAS

L. Chad

R. Nile

R. Benue

R. Niger

Key

below sea level

0–183 m

183–366 m

366–1400 m

above 1400 m

ergs

present desert boundary

orientation of 'fossil' dune systems

'fossil' drainage lines

0   600 km

interglacial, followed by another periglacial period. Quite apart from what would happen in the interglacial, the two periglacial periods would achieve much preparation of the ground for further erosion – deep freeze–thaw weathering, the production of an active layer, and so on. So the glacier acts, not on bedrock, but on bedrock weakened by periglacial activity. Taking this in conjunction with multiple glaciation, it is scarcely surprising now that glacial erosion achieves so much.

This discussion of intervening processes leads on to another idea, that of polygenetic landforms. A polygenetic landform is one that owes its origin to more than one process. It is a common phenomenon, hardly surprisingly in view of the variety of processes that must have operated in any one place during the Ice Age. Let us follow through one example, the chalk dry valley. There are lots of theories to explain chalk dry valleys. Three well known ones are that the valley was cut during a period of former high rainfall, that it was cut by rivers swollen by snowmelt during periglacial periods, and formed as hypothesised by Fagg (1923). Now in some parts of southern England it is quite likely that all three processes operated, and perhaps others, too. That being so, any one dry valley is the product of repeated phases of more than one process – it is a polygenetic landform. It is also true, of course, that one dry valley in one place need not have been formed in the same way as another dry valley somewhere else. It makes attempts to find a single hypothesis to explain all dry valleys seem rather naïve.

This in turn leads on to another idea, namely, equifinality or convergence. Equifinality is the situation where one process operating in one place produces the same landform as a different process operating somewhere else. There are lots of examples of this, too, but a continuation of the discussion of chalk dry valleys will serve very well. To judge from the debate on the origins of dry valleys, a dry valley formed in one way looks much the same as one formed in another way. So the dry valley is a 'convergent' landform. The opposite can happen too. In other words, the same process operating in two different places sometimes forms two different landforms. This is called 'divergence'. Climatic geomorphology relies on it. But that is a digression. To return to the main theme on convergence, the question now arises as to whether we can infer process from form. In other words, can we say how a landform was formed just by knowing what it looks like and exactly what shape it is? Generations of geomorphologists have done so. To take but two examples of where this approach has proved misleading, scarp-face dry valleys in southern England have been mistaken for corries, and rotational slip features have been mistaken for corries, too. So the answer to our question is: no, in view of the possibility that two different processes can form the same landform, we must be conservative and say that form alone is an inadequate guide to process.

This line of argument points to one more implication. It concerns how we define landforms. Again, one example will suffice to illustrate the generalisation. How do we define a tor? It would be easy to write something like: 'A tor is an isolated mass of jointed rock standing as a castellated mass above the general surface of a plateau, resulting from subsurface rotting of the bedrock along the joints by chemical weathering during the Tertiary followed by later stripping of the rotted rock by periglacial processes.' And that would be a mistake, not because Linton's (1955) hypothesis is necessarily wrong, but because if we found a landform looking like a tor but which was definitely formed in some other way, then we could not call it a tor. Similarly, we would suddenly have to call a particular 'tor' something else if it were demonstrated that Linton's hypothesis could not apply to it. It is all very clumsy and unnecessary. Putting to one side the massive controversy on the origin of tors, we can draw our general conclusion that tors, and landforms in general, should be defined only in terms of what they look like. The way is then clear to investigate the origin of the landform without jeopardising the use of the term. So we define in terms of form, not process; descriptively, not genetically. It is an important geomorphological principle.

Landforms as numerical evidence of climatic change

From what has been said it is obvious that some landforms are indicators that climatic change has taken place. Not so obvious, but more interesting, is the question whether any landforms are accurate measures of the degree of climatic change. In other words, can we say precisely what the climate was like when a certain landform was formed? The answer is, yes, we can, and two examples will be given, and it is left as an exercise for the reader to think of others.

One piece of such evidence is the altitude of cirques. The firn line on a cirque glacier usually occurs about three-fifths of the way between the snout and the upper limit of the ice, and the altitude of the firn line is related to the climate. It follows that for cirques now vacated by ice estimates can be made of the altitude of the former firn line and thus the former climate. The topic is discussed in detail in several textbooks, for example, in Embleton and King (1968, p. 198).

The second feature is the ice wedge. This is a recognised periglacial structure observable in present-day periglacial regions and preserved in fossil form in many mid-latitude locations, including some river terraces in southern England. Discussion of their origin is irrelevant here. The point is that it is thought that they form only when the mean annual temperature is −6°C or colder over a period of many years. Where they are preserved, therefore, they give an accurate idea of the degree of refrigeration during that cold phase.

Changes of sea level

The oscillation of climate during the Pleistocene caused worldwide changes in the relative heights of land and sea. During the glacials, sea level fell. Estimates of the amount of fall are complicated by three factors: the need to allow locally for postglacial isostatic readjustment, the fact that the glacials were not all equally cold, and the fact that there seems to have been a general fall of sea level during the Pleistocene superimposed on the oscillations caused by glacials and interglacials. In general, however, we are talking about a fall of 100–150 m. Conversely, during the interglacials sea level rose, and because the interglacials were rather warmer than at present, the sea level stood rather higher. Again, the numbers involve complications, but it seems that during the warmest part of the Hoxnian interglacial sea level stood about 35 m above present sea level, and during the Ipswichian, 15 m. To these worldwide changes must be added the local effects of postglacial isostatic readjustment. A number of landforms owe their origins at least partly to these relative changes of sea level, for example, raised beaches (Fig. 6.6), abandoned clifflines, rias (Fig. 6.7), and fjords (Fig. 6.8).

Thus far, we have been on familiar ground. Now come two lines of argument which are important but not so well known. The first concerns coastal depositional landforms; the second, river terraces.

The coast of Britain, and other areas, has a large number of depositional landforms. One tends to take that for granted, but there is a reason for such abundance, and to understand it we need to recreate the situation during one of the cold phases, say the Devensian. The sea stands about 100 m lower than today, thus exposing large areas of the continental shelf, so that much of the North Sea, English Channel and Irish Sea are dry land. Britain is under ice north of the Devensian limit, so some of the 'North Sea' and 'Irish Sea' will also be under ice. The exposed parts of the continental shelf will be the scene of extensive deposition – glacial, fluvioglacial and periglacial – and the deposits will have one thing in common: they will be easy to erode. Now picture what happens as deglaciation sets in. The sea rises, and as it does so it reworks all these deposits and pushes large quantities ahead of it. When we reach the present day with the sea at today's level, it has brought much of the material to today's coastline, forming the

Figure 6.6 A raised beach, between 6 and 17 m above present sea level, at Portland, Dorset (from Brunsden & Doornkamp 1977).

multitude of depositional landforms such as spits, bars and beaches which are so familiar. So the abundance of such features is a consequence of the effects of climatic change in conjunction with the presence of an extensive continental shelf beyond our shores. This is true for specific landforms, of course, not just deposits in general, so the sequence of events described above is thought to explain the origin of, say, Chesil Beach (Fig. 6.9) as described in detail by King (1972, pp. 307–11) and summarised in Sawyer's (1975, pp. 59-61) well known A-level textbook. The logic of the argument above would also seem to suggest that coastal deposits would not occur in vast quantities all the time, but only after certain combinations of events. They might have been quite rare at certain times in the geological past. The whole discussion adds a new dimension to the familiar dictum that coastal depositional landforms are ephemeral, especially since the sea has been roughly at its present level for only about 6000 years.

Figure 6.7 The ria at Solva, Pembrokeshire: a valley drowned by the postglacial rise of sea level (after Kay: from Small 1978).

Some modern thinking on river terraces will serve as a final example of the impact of climatic change. The traditional view is that terraces are old floodplains that have been left stranded at high levels by the incision of the river. The incision is due to a fall in sea level, itself due to a cold phase. If so, the material of which the terrace is made – we will call them terrace gravels – must have accumulated during interglacial conditions, or, more exactly, to get a rising sea level, early interglacial conditions. Now a number of analytical techniques can be brought to bear on the problem of the depositional environment of terraces, so this theory can be tested. It turns out that some terraces were deposited under interglacial conditions, so they fit the theory well, but some were not. Indeed, there is no single set of conditions that led to the deposition of all river terraces, and it is clear that some were deposited under periglacial conditions. The Summertown–Radley terrace of the Thames near Oxford is one of many examples, and the fossil ice wedge in the terrace gravels mentioned earlier is just one of the pieces of evidence (Fig. 6.10). That leaves us with the problem of

Figure 6.8 A fjord: Milford Sound, New Zealand (Aerofilms photograph).

Figure 6.9 Chesil Beach from Portland (from Sawyer 1975).

Figure 6.10 An ice wedge pseudomorph from the Stanton Harcourt exposure of the Summertown–Radley terrace, with an upper layer of convolutions (from Smith & Scargill 1975).

how terraces can accumulate under periglacial conditions with a low and perhaps even falling sea level.

We must begin by questioning the assumption that terraces are associated with sea-level change. The association seems obvious but that does not necessarily mean it is right. The point is that, because of the existence of the continental shelf, as the sea falls through 100 m it also retreats across the continental shelf by several tens of kilometres. So rivers flowing into the sea have their courses lengthened; they flow across the continental shelf to the new coastline. Now the gradient of the continental shelf is very gentle, comparable with the lower reaches of most large rivers. Therefore, the long-profile of the river remains 'graded'. There is no reason to suppose that the river will incise into its own floodplain.

Having freed the discussion from the complications of sea-level change, we can now concentrate on explaining how terrace gravels can accumulate under periglacial conditions. Any answer is bound to be speculative, but one possibility is that since the periglacial surface is so easily eroded during spring and summer, streams in flood due to the spring snowmelt will acquire an enormous load. On encountering gentler gradients lower down the main stream, much of this load is deposited. In fact, the discussion by no means ends there since there is much of relevance that has not been introduced, but enough has been said to show how applying our knowledge of Pleistocene climatic change forces us to rethink some geomorphological issues.

Conclusion

This chapter has had three main themes. First, Pleistocene geomorphology is a vast and active field of current research, and much is known about the events of the period. Secondly, that knowledge has many geomorphological implications. Thirdly, and perhaps most importantly, the research is revealing the *chronology* of the Pleistocene: this is the last, but vast, vestige of the historical approach, the search for sequences of events. Apart from that, the two disciplines of pure and applied geomorphology have developed along quite different lines, and these are described respectively in Parts III and IV.

PART III

Modern pure geomorphology

7 Form

Introduction

By 'form' we mean the physical dimensions of a landform or area of relief – its size and shape. Accurate measurement of form must be one of the most fundamental tasks of geomorphology. In the last analysis geomorphology is concerned with land form. Everything else depends on knowledge of form.

The various dimensions of a landscape or of individual landforms can be measured, and any relationships between them can be investigated. This branch of geomorphology is called morphometry. Sometimes accurate measurement of land form reveals a close approximation to a recognised mathematical shape such as a parabola or a logarithmic curve. Studies of form also include some very small features, and the chapter concludes with a discussion of some of these.

Morphometry

Any significant or relevant measurement of the size and shape of a landform is called a 'parameter'. Therefore, height, length, width, angle and so on are all parameters. We are concerned in this chapter with making these measurements and with relationships between the parameters. There are important quantitative and statistical relationships between some parameters, and in studying these relationships – called the 'laws of morphometry' – we are in effect studying morphological systems. These are considered further in Chapter 10.

Since measurement of form is relevant to all geomorphology it follows that morphometric studies can be carried out in all branches of the subject. Indeed, the term 'general geomorphometry' is sometimes reserved for quantitative analysis of the relief of an area without necessarily referring to any specific landforms (see, for example, Evans 1972). In this discussion, however, I am going to concentrate just on four well known examples: drainage basin, slope, beach, and pebble morphometry.

Drainage basin morphometry originates with the work of Robert E. Horton (1932, 1945). In Horton we come again to one of the great men in the history of the development of geomorphology. He was an American, and by profession an engineer, not an academic. He is described as '. . . an engineer of unique talents . . .' by Leopold,

Table 7.1 Sixteen morphometric properties of drainage basins.

1 perimeter of the drainage basin (p)
2 area of the basin (a)
3 maximum elevation within the basin (H_{max})
4 minimum elevation within the basin (z)
5 relief of the basin ($H_{max} - z = r$)
6 relative relief of the basin ($100r/p5280 = R$)
7 circularity of the basin (C)
8 total length of streams of order u (l_u)
9 total length of all streams (Σl_u)
10 drainage density ($\Sigma l_u/a = D$)
11 number of stream channels of order u (n_u)
12 total number of all stream channels (Σn_u)
13 stream frequency ($\Sigma n_u/a = F$)
14 ruggedness number ($Dr/5280 = H$)
15 bifurcation ratio ($n_u/n_{u+1} = R_b$)
16 hypsometric integral,† $I = \int_0^{100} da/dr$

†For a working definition of the hypsometric integral, see Strahler (1952).

Table 7.2 Some morphometric properties of slopes.

1 maximum (β) angle of slope
2 height of profile (Ht)
3 length of profile (Lg)
4 average angle of slope ($Ht/L_g = \sin \alpha$)
5 gradient of stream channel at base of slope (γ)
6 depth of soil or weathered mantle (W)
7 principal grain-size characteristics of the soil or mantle (e.g. D_{10})
8 pH value of the soil or mantle
9 moisture content of the soil or mantle, as a percentage of its dry weight ($S\mu$)
10 porosity or void ratio of the soil or mantle (V_R)
11 organic content of the soil or mantle, by weight (W_0)
12 root weight of the soil or mantle (W_R)
13 degree of surface cover by plants (S_c) in percent
14 average height of the vegetation (H_v)

Table 7.3 Some morphometric properties of beaches.

1 maximum angle of slope (θ)
2 length of the profile (Lg)
3 height of the profile (Ht)
4 average angle of the slope ($Ht/Lg = \sin \alpha$)
5 length of the maximum angle section (L_{sm})
6 proportion of the total profile length occupied by the maximum angle section ($L_{sm} \times 100/Lg = SM_L$)
7 height/length integral
8 average D_{84}, D_{50} and D_{10} grain sizes of the material on the maximum-angle section
9 average porosity or void ratio of the material on the maximum-angle section (V_R)
10 average penetrability of the material on the maximum-angle section (P)

Wolman and Miller (1964, p. 134). As an engineer, his approach was essentially practical. As early as 1919 he was studying interception in the forests of the United States, and in 1924 he contributed an article on the design of stormwater drains. He concerned himself with many aspects of geomorphology and hydrology, including soil erosion and dustbowl problems in the 1930s, an article on infiltration capacity (1933), and the concept of overland flow which offered an explanatory model of runoff erosion. However, it is his pioneer work on quantitative drainage basin analysis that secures his place in history. In fact, his laws of stream numbers and stream lengths are sometimes called the laws of morphometry, but this is a poor use of the term because, strictly speaking, they are not laws and because morphometry implies much more than this. Indeed, there are at least 16 measurements that might form the basis of morphometric analysis of drainage basins. They are listed in Table 7.1. Some of these parameters are related to each other, thus forming the basis for the laws of drainage basin morphometry. It must be stressed that some of these measurements constitute major areas of enquiry in themselves. For example, there are a number of indices designed to express the plan shape of a drainage basin – the circularity ratio is by no means the only one – and hypsometric analysis is a well defined branch of the subject discussed by McCullagh (1978).

In the same sort of way, measurements of slopes (Table 7.2) and beaches (Table 7.3) allow morphometric analysis of those landforms. However, morphometric analysis can be applied at a completely different scale, and pebble morphometry provides a good example of this. The idea behind this is that pebbles from different environments (e.g. coastal, glacial, fluvioglacial, aeolian, etc.) have different morphometric properties, and this will assist in assigning an origin to fossil deposits. There is a surprisingly large number of indices for expressing the shape of a pebble, some of them extremely complex, but a long-established and simple one is Cailleux's Flattening Index

$$F = (L + l)/2E$$

where L = major axis (length) of pebble, l = width, and E = breadth.

Morphometric work is tedious and time consuming. The justification for it lies in four factors:

(a) It is the only way to give precision to description of landform.
(b) It facilitates the comparison of one landform with another, especially in the case where one has not actually seen one of the landforms.
(c) It makes possible correlation between form and other variables that one can measure, for example, between form and process.
(d) It can be used for prediction. This is important because it is a

practical application. For example, a relationship between drainage basin morphometry and the shape of a flood hydrograph enables one to be more precise about the magnitude and timing of a flood peak.

Mathematical shapes

A rather different result of the precise measurement of form is that sometimes a landform turns out not to be the shape that everyone thought it was, or that a landform conforms roughly to a recognised mathematical shape. It must be remembered, of course, that in reality a given landform might display a variety of shapes in different places, and that the resemblance to a mathematical form is rarely exact. Nevertheless, the 'average' shape is often near enough to a mathematical shape to be interesting.

A classic example of landforms turning out, on measurement, to have unexpected shapes is the form of hillslopes in England. These were assumed to be convexo-concave, as Davis taught, but this was evidently a case of the observer seeing what he expected to see, for measurement later showed that the straight, rectilinear slope element is common in England. Another example, discussed by Small (1978, pp. 357–8), is that U-shaped glaciated valleys in fact display a wide variety of forms and often are not U-shaped (see Fig. 7.1). Valley cross-profiles are also examples of approximation to mathematical curves, for it turns out that glaciated valleys are often parabola-shaped (Fig. 7.2), that is, they conform roughly to the equation $y = ax^2$. All this calls into question the long-held belief that glaciated valleys have flat floors. Another example from glaciation is that the plan form of drumlins has been shown to resemble a lemniscate loop (which is a mathematical expression of the streamlined form) and an ellipsoid (each contour is approximately an ellipse). This topic is discussed further by Embleton and King (1968, pp. 322–4).

Probably the best-known example of resemblance to a mathematical shape is the accordance of the long-profiles of rivers, or at least segments of such profiles, to a logarithmic curve of the form $y = k(a - \log x)$. This one 'discovery' has led to a massive literature, much of it not to the credit of geomorphology, on a variety of related subjects. These include attempts to show that streams become graded when they assume the smoothly concave shape, efforts to reconstruct former sea-levels from partial profiles, and hypotheses to explain the smooth logarithmic shape.

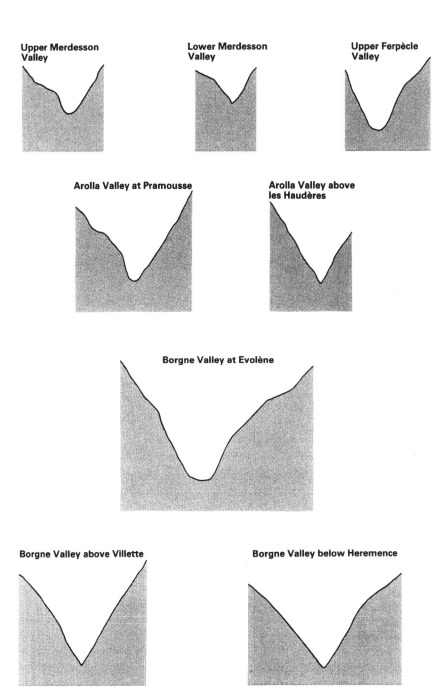

Figure 7.1 Cross-profiles of the Val d'Hérens (from Small 1978).

Figure 7.2 Glen Rosa, Isle of Arran: a view northwards towards Cir Mhor (left) and the col leading into the head of Glen Sannox (after Kay: from Small 1978).

Small-scale forms

A final point to be made on the subject of studies of form is the great range of scale with which the geomorphologist concerns himself. In particular it should be noted that some very small forms are regarded as legitimate objects of study. They often represent the detailed configuration of the surface of a landform which is studied in its own right. For example, small ribbons and ripples are studied on desert sand dunes, clints and grikes on limestone pavements, and terracettes on slopes. Sometimes these small-scale landforms, by virtue of their size, lend themselves to laboratory simulation. This is the case, for instance, with the bedforms on rivers – sand waves, dunes and antidunes (Gregory & Walling 1973, p. 163).

One branch of geomorphology where much work has been done at this scale is glacial and fluvioglacial erosion. The features produced include striations, grooves, polished surfaces, friction cracks, cavetto forms, *Sichelwannen*, curved and winding channels, bowls and potholes, and plucked surfaces, and were treated as the subject of a whole chapter by Embleton and King (1968).

There have also been many studies of small-scale forms on beaches. The forms are:

(a) *Beach ridges*, or berms, of which there may be up to six and which are built up by constructive waves at various levels.

(b) *Patterns of ripples* which are well developed on the lower beach when this is formed of sand. Symmetrical ripples result from wave action, asymmetrical ripples from tidal currents.

(c) *Channels.* Minute and often anastomosing channels in the sand.

(d) *Ridges and runnels,* which are broad and gentle rises and depressions aligned parallel to the shoreline.

(e) *Cusps.* These are small regular embayments developed on the face of shingle or at the junction of a shingle beach and a sandy beach. The re-entrants are floored by sand and small shingle while the seaward projecting horns are of large pebbles. Cusps have been studied in some detail, and although their origin is unclear, it is easy to see how, once initiated, they persist – each wave washes most strongly into the hollow, carrying larger pebbles to either side and leaving the depression floored by finer particles.

Conclusion

The study of form surely lies at the centre of geomorphology. The subject exists because generations of academics have had their interest aroused by the shape of the surface of the Earth. The present generation adopts a quantitative approach to this, measuring form accurately and searching for statistical relationships between the various parameters. But, of course, the enquiry does not stop there. One wants to know why the landform is there and why it is the shape it is. So in the next chapter we pass on naturally to a consideration of the processes that create landforms.

8 Process

Introduction

The study of present-day processes is almost certainly the most important single theme in modern geomorphology, for it lies at the core of pure geomorphology and indeed of applied studies as well. However, this has not always been so. In Part I we saw that the early lead in process work given by Gilbert and others was not followed as geomorphology became obsessed with the historical approach. There were exceptions, and one can cite work on glaciers (e.g. Johnson 1904, Gilbert 1906, Thorarinsson 1939), on shoreline processes (e.g. Gaillard 1904, Johnson 1919, Lewis 1931) and on aeolian processes (Bagnold 1941) to show that some important process work was being done during the first half of this century. Nevertheless, the study of contemporary processes by geomorphologists was relatively un-important during these decades, and analysis of fluvial processes was left to hydrology and hydraulics (e.g. Linsley *et al.*, 1949). It was not until about 1960 that this began to change, and it is probably fair to quote Leopold *et al.* (1964) as the principal pioneer work on modern process studies.

Development of process geomorphology

There are many reasons for the rise of process geomorphology. It was recognised that Davis, although attributing landscape development to structure, process and stage, himself failed to emphasise process, and that the historical approach in general had suppressed the importance of the dynamics of present-day processes. There was also a desire to make geomorphology less of a descriptive science and more of an explanatory one, and explanation of landforms can come only from an understanding of process. This direct link between creative process and resultant form is now recognised as process–form studies and is analysed in the context of process–response systems. The wish to give the subject a greater explanatory content is also related to the contemporary development of applied geomorphology, since to be of practical value explanations must exist and must be correct. The related topic of assessing the importance of man as a geomorphic agent also relies on an understanding of process. A final point is the parallel development from the 1950s onwards of quantification in geomorphology. It is too glib to say that process studies lend themselves to quantification: often measurement throws real light on the nature of a process. An early example of this was the demonstration by Leopold and Maddock (1953) that, in general, the mean velocity of a river does not, as seems obvious, decrease in a downstream direction.

The importance of process studies today is such that all modern textbooks use process–form relationships as their theme, and in particular it is a sign of the times that there is a number of books dealing specifically with the processes, for example, Chorley (1969), Weyman (1975), Weyman and Weyman (1977), Derbyshire *et al.* (1979) and Embleton and Thornes (1979).

The main aspects of process geomorphology will now be reviewed. However, it must be remembered that process studies are by no means without difficulties. Fieldwork can be awkward, with processes operating with a duration and an intensity that make their study problematical. Processes in nature act in combination; how does one separate out one for analysis? The link between process and form is not always obvious; form can influence process. And, possibly most important, although we know what happens in general terms and can observe the results of process, our knowledge of the detailed mechanism of many processes is not complete. As will be seen shortly, this is a field in which geomorphologists have been busy.

Weathering, biological processes and soils

The various recent textbooks on weathering, such as Ollier (1969) and Carroll (1970), adopt slightly different approaches to the topic, but a useful classification of weathering processes is as follows.

(a) Processes of disintegration:
  (i)   Crystallisation processes – salt weathering and frost weathering.
  (ii) Temperature-change processes – insolation, exfoliation.
  (iii) Weathering by wetting and drying.
  (iv) Organic processes such as root wedging and colloidal plucking.

(b) Processes of decomposition:
  (i)   Hydration and hydrolysis.
  (ii) Oxidation and reduction.
  (iii) Solution and carbonation.
  (iv) Chelation.
  (v)   Biological chemical changes.

For the most part this will be a familiar list, but for readers who are keen to see the modern developments certain points are worth stressing. First is the great significance accorded to crystallisation processes. They are generally the most important agents in the natural weathering of building stones in Britain. Second is chelation, which involves the holding of an ion, usually a metal, within a ring structure of organic origin. Chelating agents can extract ions in otherwise insoluble solids, and enable the transfer of ions in chemical environments where they would normally be precipitated.

The third important modern development is the emphasis on biological processes. Traditionally this has been something of a neglected topic in British geomorphology, although it must be said that the climatic geomorphology school has always appreciated their role. Under different circumstances biological processes can have the effect of either accelerating or retarding both mechanical and chemical weathering. The main effects are as follows.

(a) *Simple breaking of particles*, as by the eating or burrowing of animals, or by tree falls and the pressure exerted by growing roots.

(b) *Transfer and mixing of materials*, mainly by animals, moving mineral materials into areas of different weathering effects.

(c) *Simple chemical effects*, as when solution is enhanced by the carbon dioxide produced by respiration.

(d) *Complex chemical effects* such as chelation, and the formation of complexes of organic-mineral substances.

(e) *Effects on soil moisture*. These effects are partly due to the water-holding properties of root masses and humus, and partly to the shade effects of plants.

(f) *Effect on ground temperature*, by shade, by production of heat as in fermentation, and by moving material to and from the surface.

(g) *Effects on pH of surfaces*. These are due both to respiration and direct chemical effects.

(h) *Protection from erosion*, both water erosion and wind erosion, causes less exposure of new surfaces and therefore less total weathering under most circumstances.

Soils are closely related to weathering. It is sometimes said that there is little difference between them, and certainly soils are a product of weathering. Soils are also related to biological processes, being the zone in which many of them occur. The geomorphological significance of soils has long been recognised, although the full extent of the common ground between them has been realised only in recent years (Birkeland 1974, Gerrard 1981). Where they occur, soils are the zone of contact between rock and subaerial processes; geomorphology 'happens' here. It follows that recent interest in the properties of materials (Ch. 9) includes the properties of soil. Certain soil properties have immediate geomorphological implications, for example, cohesion, porosity and infiltration capacity. Some movements of soil masses, such as soil creep, are geomorphologically important, and the behaviour of soil is closely linked to slope studies. The close relationship between soils and landforms is expressed by the catena concept, and it is interesting that some of the effects of soils on landform are indirect, such as the effect via natural vegetation and the effect that soil type has on man's use of land. Soils also enter into applied geomorphology, soil erosion, for example, being a recognised geomorphological hazard; this aspect will be considered further in Chapter 12.

Erosion

Rightly or wrongly, this is the group of processes which has dominated geomorphological thinking over the years. The forces of construction originating within the Earth, such as mountain-building and vulcan-icity, are of marginal interest to the geomorphologist, and he has concentrated instead on the processes eroding the landmasses and the landforms so produced. As this chapter shows, the interest of geomorphologists today is by no means confined to erosion, but nevertheless there have been some interesting new developments in the study of erosional processes.

The main erosional processes should be well known. Rivers erode by abrasion (usually regarded as the most important one), corrosion and hydraulic action, there being two types of hydraulic action – evorsion (the direct impact of water) and cavitation (pressure effect under very high-velocity flow). To this list of processes must be added entrainment, whereby particles on the bed and banks of the river become incorporated into the suspended load. It is here that the dividing line between erosion and transport becomes very thin. The main glacial processes are plucking – quantitatively the more important one – and abrasion. However, the distinction between cold and temperate glaciers is undoubtedly important, because temperate glaciers are at pressure melting-point at their soles and are therefore geomorphologically active, whereas cold glaciers are frozen to bedrock, at least partly, and are therefore incapable of either plucking or abrasion. Also important in cold environments is nival erosion, which results from the presence of a snowpatch in a hollow, and which is an important periglacial process and almost certainly relevant to the

91

initiation of cirques. In recent years there has been much rethinking of desert processes. More emphasis than hitherto is placed on the role of chemical weathering in the formation of, for example, deflation hollows; and on the role of contemporary fluvial erosion in the form of flash floods and sheetflow. At the same time the significance of abrasion has been played down, this being confined to the zone of saltation, usually within about 1 m of the ground.

Two important general developments in the study of erosional processes are worthy of note. The first is the recognition that these processes rarely act on unweathered and unprepared material. This has had the effect of encouraging research on the preliminary processes. We have already looked at an example of this in Chapter 6, where it was shown that glacial erosion is by no means just a matter of abrasion and plucking. Another factor involved in the same situation is the pressure-release mechanism. As the glaciated valley becomes deepened and the ice thickens, rock is being replaced by ice, which has a lower density. The bedrock responds to the decreased pressure by expanding at the surface, creating pressure-release fracturing in the rock. Further erosion is thus assisted. This situation – whereby erosional processes involve things that one does not think of straightaway, and work on preprepared materials – is repeated throughout geomorphology.

The second new development concerns the detailed operation of erosional processes. In response to our lack of knowledge of exactly how processes work, research workers concentrate now on the nature of fluid motion, on energy, on forces, on resistances, and on the properties of materials. Such topics are considered in detail, for example, by Embleton and Thornes (1979) on geomorphological processes in general, by Finlayson and Statham (1980) on hillslope processes, and by Kirkby and Morgan (1981) on soil erosion. The mechanics of erosion are very closely related to the properties of materials and the subject is discussed further in the next chapter.

Transport

Most students probably do not think of transport as a separate process at all. They are taught the difference between capacity and competence, and that transport is made up of suspension, solution, saltation and various types of creep, rolling and sliding, but otherwise transport becomes the link between erosion in one place and deposition in another that one takes for granted. In fact, however, transport is a well researched phenomenon, and it is of great practical importance in, for example, soil erosion by wind and water (see, example, Statham 1977).

The classic work on transport by streams was done by Hjülstrom (1935). Figure 8.1 appears in many standard textbooks. The curve of mean velocity indicates the stream velocity needed to entrain (move into suspension) a variety of particle sizes. The mean fall velocity curve shows, in a similar way, the circumstances of deposition. Two points are noteworthy. First, sand (0.06–2.0 mm) is more easily moved than both larger- and smaller-size fractions. The fine-grained particles, silt and clay, are more cohesive. Secondly, the velocity required to entrain material is higher than that needed to transport it. This applies particularly to the smallest grain sizes which, once entrained, can be maintained in suspension at extremely low velocities. This helps to explain, among other things, the suspended load of a river, beach cusps, and the absence of dunes on loess deposits.

A rather different comment that must be made about transport is that there is a modern view held by some geomorphologists that some agents of erosion might be better interpreted as agents of transport. For example, we are all taught at school that rivers erode by attacking their bed and banks. That is true, but it is most unlikely that rivers acquire all their load in this way, and it might be more accurate to say that a river is an avenue of transport evacuating material supplied to it by slope processes. It is just a different way of looking at things. A similar sort of situation arises at the coast where a non-resistant and incoherent rock meets the sea. For example, just north of Weymouth in Dorset, Furzy Cliff is cut in Oxford Clay and exhibits much slumping and slipping (Fig. 8.2). Here it is probably less true to say that the sea is eroding the cliff and more true to say that the sea is acting as an

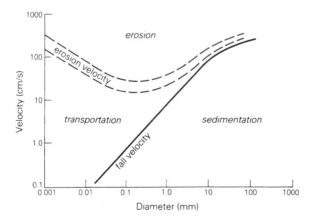

Figure 8.1 The relationship between the velocity of stream water, and the erosion, transportation and deposition of particles of different size (logarithmic scale) (from Morisawa 1968).

Figure 8.2 Furzy Cliff, north of Weymouth (from Sawyer 1975).

agent of transport removing debris supplied to it by subaerial and slope processes.

Slopes and mass movements

In that the landscape is all slopes, there is nothing more central to geomorphology than slope form and process. However, the long debate as to whether slopes decline or maintain a constant angle as time passes – wearing down or wearing back – has fallen into disfavour, and modern work on slopes concentrates almost exclusively on present-day forms and processes. As we shall see in Part IV, much of this work is of great practical significance, involving as it does matters of ground subsidence and slope instability. Most modern textbooks on geomorphology reflect recent developments and treat slopes dynamically as open systems. The main textbooks dealing specifically with slopes are those by Brunsden (1971), Carson and Kirkby (1972), Young (1972), Schumm and Mosley (1973) and Finlayson and Statham (1980).

There is more than one way of classifying slope processes, and every textbook seems to employ a slightly different classification and even a slightly different terminology. The two main groups of processes are those involving the action of water, and mass movements. However, the distinction is not as clear as that, for there is a continuum of rock–water substances all the way from rockfalls at one end to streams at the other. Somewhere in the middle a muddy waterflow and a runny mudflow grade imperceptibly into each other. Looked at in this way, slope processes can be regarded as transport systems.

The three main types of process involving the action of water are raindrop impact (producing splash erosion), overland flow (as distinct from channel flow) and subsurface flow, of which throughflow is probably the most important. Concentrations of throughflow, called 'piping', can lead to the heads of surface gullies and help to extend the gullies by eroding subsurface material (Gilman & Newson 1983, Jones 1983).

Mass movements are the result of shear stresses on slopes (caused by gravity, the weight of material and soil water) overcoming the inherent resistance of the materials (made up of the cohesive properties of slope particles and their internal friction). Mass movements themselves can be classified in a number of ways. A simple classification is into flows, slides and falls. Some of the flows are slow, such as soil creep, talus creep, rock creep and rock glacier creep. Others are rapid, such as mudflows, earthflows and solifluction. The slides include different types of rock and debris fall. One thing is now very clear – mass movements are extremely common in Britain. In almost any part of the country, one can scarcely begin a description of the morphology of hillslopes without encountering some sort of mass movement. Interestingly, some of them are fossil features inherited from late-Pleistocene periglacial periods, and others result from the fact that oversteepened slopes produced during the Pleistocene are not stable under present-day conditions.

Deposition, depositional landforms and sediments

There is little doubt that, traditionally, geomorphologists have paid rather more attention to erosional landforms than to depositional features. A study of almost any textbook would reveal the imbalance. This is undoubtedly a failing of the subject, for it is clear that material eroded from one place will be deposited in another, so it is obvious that the two sets of features have equal merit as objects of study. There are probably several reasons for the relative neglect of depositional landforms. Davis emphasised erosion processes in landscape sculpture, wrote mainly about erosional landforms, and called his cycle the 'cycle of erosion'. Denudation chronology focused on erosion surfaces and the types of erosion that might have formed them. In general, erosional landforms are more spectacular to look at and invoke processes that are interesting and exciting to debate. To most young geographers, the form and origin of corries and arêtes at one end of a glacier (Fig. 8.3) are altogether more attractive propositions than the moraines at the other (Fig. 8.4).

Over the past 20 years or so geomorphology has gone a long way towards correcting the imbalance. Apart from the more prominent position given to deposition in most recent textbooks, a number of works have dealt almost exclusively with deposition, for example, Allen (1970), Hails and Carr (1975), Komar (1976) and Briggs (1977). The rise of depositional studies can be attributed to various factors in addition to geomorphologists simply making up ground in what was recognised as a neglected area in process geomorphology.

The first point is that, in general, geomorphologists have a much better understanding of the origin of depositional landforms than of erosional features. This is because, by definition, the material making up the depositional landform is there in the landscape for them to observe and analyse, whereas with erosional features all they have is the form of the feature, so they either have to infer process from form or observe the process actually forming the feature somewhere today. Further, the material making up a deposit can be analysed using a host

Figure 8.3 Snowdon (Aerofilms photograph).

Figure 8.4 An end moraine (from Brunsden & Doornkamp 1977).

of field and laboratory techniques so that the nature of the depositional environment can be recreated.

The second point is that it is often possible to date a deposit with some measure of certainty using, for example, radiocarbon or uranium-series dating. Erosional landforms can virtually never be dated in this way. Indeed, our knowledge of Pleistocene chronology and of the role of periglacial processes in this country rests almost entirely on analysis of deposits and depositional landforms.

A third point is that the efficacy of erosion as a process is often inferred indirectly from measurements of sediment. For example, the erosional work performed by a river can be estimated by measuring the load it is carrying, the rate of denudation of an area is reflected in the build-up of silt in a reservoir, and the erosional abilities of glaciers can be gleaned, to some extent, from the load carried by streams issuing from the glacier.

The final consideration is that some depositional studies are of immense practical significance and have a place in applied geomorphology. The build-up of silt in reservoirs, river channels and harbours is a case in point. Another example concerns beach nourishment: this succeeds only if near-shore sediment dynamics are well understood.

Rates of present-day processes

It may well have occurred to the reader that one result of these many process studies should be that we know something of the numerical rate of operation of present-day processes, and this is indeed the case. Results are available from several different geomorphological environments. For example, around the coast of Britain, average rates of cliff retreat range from less than 2 cm per year on some hard rocks in the North West to about 30 cm a year on Chalk, about 1 m a year on Tertiary sands and clays, and 1.5–2.0 m a year on the boulder clays of Holderness. Much is known about the rate of erosion of limestone from many parts of the world, and indeed there is a long-standing debate in karst geomorphology as to whether variations from place to place should be attributed to lithological or climatic factors. Most results seem to indicate a rate of surface lowering in the range 1–3 cm/1000 years, although whether this is the best index for limestone areas is open to question. Quantitative aspects of glacial erosion are considered by Embleton and King (1968), a discussion quantifying the well known assumption that glacial erosion is several times more potent than fluvial erosion.

Attempts to establish rates of erosion over very extensive areas, such as whole climatic regions, for instance, are fraught with difficulties.

The rates vary considerably from time to time, and from place to place at the same time; it is difficult to judge the extent to which man is affecting natural rates; and even the rates themselves are in doubt because different lines of evidence often do not produce comparable results even for one area. Given the proviso, therefore, that we are at best talking only about orders of magnitude, it looks as though the world average rate of denudation is about 3 cm/1000 years, ranging from about 1 cm/1000 years in hot desert lowlands to about 100 cm/1000 years over tropical monsoon mountains and currently glaciated mountains. The evidence for many different environments is discussed by Rice (1977, Ch. 10).

There are certain implications of this kind of investigation. One is that it becomes possible to make classifications based on rates of present-day processes. For example, we can identify the Icelandic-type periglacial environment as an active geomorphological environment in contrast to, say, the very arid environment where rates of process are low. Davies (1980) has used rates of contemporary processes to classify world coastlines into high-energy and low-energy environments. Working at a very high level of generalisation at a world scale, he classifies mid-latitude coasts as high-energy, and low- and high-latitude coasts as low-energy environments. Local variations can be just as great, however. In Britain, for example, we have the full range from high-energy cliff and storm-beach coastlines to low-energy saltmarsh environments.

Another implication concerns the magnitude and frequency of geomorphic processes that are of greatest importance in the long run. That needs to be made clear. When a river floods, an enormous amount of erosion takes place. First thoughts, then, would seem to suggest that most geomorphic work on rivers is done during floods. But rivers do not flood very often. It may be that much lower flows, less spectacular but occurring more often, might achieve more erosion in the long term. We can express the problem as an exact question: What magnitude and frequency of river flows are of greatest geomorphological significance overall? The answer was established by Wolman and Miller (1960) who, using the term 'the frequency concept' to express their ideas, showed that in the long run most geomorphic work is done by flows of more moderate magnitude and frequency than large floods – those occurring about once or twice a year.

Tectonic processes and plate tectonics

It is a curious feature of process geomorphology that tectonic processes are not normally thought of as process studies at all. This is reflected in their exclusion from modern books on processes such as Weyman and Weyman (1977) and Embleton and Thornes (1979), and there is little doubt that most geomorphologists think along these lines. It is doubtful if there is any particular reason for this. It is just a matter of how terminology comes to be used by academics. Process study has come to mean subaerial process study, but there is no logical reason why it should not include tectonic processes as well, and that is the reason for discussing them here.

There is considerable current interest in the role of tectonic processes in landform development, and the subject is extensively treated in two recent textbooks: Weyman (1981) and Ollier (1981). The topic is just one aspect of what has come to be known as structural geomorphology (Ch. 5), and the whole field of study has received a great boost during the past 20 years from the emergence of the concept of plate tectonics. Plate tectonics has not only provided the explanatory framework for tectonic processes, but it has assumed the role of a paradigm in geology, as discussed in Chapter 15.

Plate tectonics is a grand and spectacular integration of existing ideas, the most important of which are continental drift (Tarling & Tarling 1973) and sea-floor spreading (Vine & Hess 1970). There is an enormous literature on the subject, and quite apart from summaries that appear in many modern textbooks on general geomorphology or general geology, and quite apart from countless magazine articles, the list of textbooks specifically about plate tectonics is impressively lengthy: Runcorn et al. (1962), Calder (1972), Gass et al. (1972, and especially Ch. 19 by Oxburgh), Le Pichon et al. (1973), Hallam (1973), Wilson (1977) and Heather (1981). I do not propose to explain the concept of plate tectonics here, nor to show how it helps to explain the major relief features of the Earth. All this will be familiar to most, if not all, readers, and if it is not then there is ample material for the reader to consult.

What I do want to do is discuss the implications of plate tectonics for geomorphology. From a purely geomorphological point of view this is by far the most important aspect of plate tectonics, yet it is one which has had surprisingly little airing in the literature so far. The list is by no means obvious, because although plate tectonics occupies an absolutely central position in geology, there is a difference between geology and geomorphology, and the implications specifically for geomorphology do need to be thought out. They are as follows:

(a) Plate tectonics offers a comprehensive and totally integrated explanation for the major relief features of the Earth and for a number of specific landforms which the geomorphologist, by tradition, has included in his field of study: mountain ranges, block mountains, basins, rift valleys, volcanoes, island arcs and ocean trenches.

(b) In the same way, plate tectonics is concerned with a number of other concepts and phenomena that are important in geomorphology. The phenomena include folding, faulting, earthquakes and isostasy, all of which can be important in landform development. The concepts include the idea that tectonic forces, in uplifting land, are a major provider of energy into landform systems, and the idea that in some parts of the world tectonic movements have substantially altered the relief and drainage of an area (see, for example, Doornkamp's (1977) discussion of East Africa).

(c) Plate tectonics has a number of practical applications. The subject of applied geomorphology will be discussed in more detail in later chapters, but for the moment we can note that knowledge of plate tectonics theory helps geomorphologists to predict and even possibly control certain geomorphological hazards such as volcanoes and earthquakes. Plate tectonics is also important in the evaluation of resources. Many of the principal ore deposits are associated with rising convection currents from within the Earth. For example, as the African and Arabian land masses separate along the line of the Red Sea, sizeable deposits of many ores are being created. These include gold and silver deposits, as well as invaluable sources of copper, iron and other metals. From this it follows also that knowledge of ancient plate movements might lead to the discovery of deposits of ore in areas that are not now active. The implications for search procedures are obvious.

(d) Plate tectonics has highlighted certain points about the significance of scale in geomorphology. Scale – like distance, direction and time – is one of the most basic concepts in geography, and yet we are not always aware of it. In this respect, incidentally, King's (1980) recent book is interesting in that she uses scale as the theme for organising the text. However, that is a digression. The point is that in geomorphology one is not always working at a scale at which plate tectonics is significant. Let us take an example to illustrate this. In drainage basin studies, much geomorphological work is done on small catchments, such as in the Mendips (see, for example, Weyman & Weyman 1977, Ch. 2). What are the implications of plate tectonics for such studies? None. So what scale of drainage basin do we need to study so that plate tectonics is relevant? Something like the Amazon. Now what geomorphological work has been done on the whole of the Amazon basin? Almost certainly none. The scale is wrong.

However, that is not to say that plate tectonics is always unimportant at the local and regional scale. Increasing numbers of microplates and plate plasterings are being recognised, and the structural movements exercise an important control over various aspects of marine geomorphology: coastal form, erosion rates, guyots, coral reefs, seamounts, volcanic plumes, and others.

(e) It helps to reaffirm that geomorphology is the study of the landforms, materials and processes of the whole Earth, not just the land portion of the planet. It is easy to forget the ocean floors. A unifying theory like plate tectonics helps one to remember.

(f) It forces academics to analyse the relationship between geomorphology and geology. Plate tectonics is central to geology. It also has implications for geomorphology. That presumably has the effect of drawing geomorphology closer to geology. We are benefiting from their windfall. It is all very arguable as to whether geomorphology is a branch of geology or whether both exist separately as disciplines within the earth sciences, but either way it seems likely that plate tectonics has strengthened the bonds between the two subjects. And yet, in another way, exactly the reverse is true. Vast parts of geomorphology make little or no use of plate tectonics. The revolution has passed them by. That would seem to suggest that plate tectonics has had little influence on geomorphology in total. A possible implication of all this is that geomorphology might 'lose' structural geomorphology to the earth sciences. On the other hand, perhaps it is just another example of the traditional interdisciplinary nature of geomorphology, the way certain branches of the subject make use of certain ideas from certain other disciplines. It is all food for thought.

Conclusion

Process studies have risen to a position of pre-eminence in the subject, and indeed to many modern students process geomorphology *is* geomorphology. Process is more than just erosion. The field also includes weathering and biological processes, transport, deposition, and tectonic processes. Some of these have not always received appropriate emphasis from British geomorphologists. The current state of knowledge of the rates of operation of processes, although incomplete, is sufficient for us to be able to classify the major world environments on the basis of geomorphological activity.

9 Materials

Introduction

The idea that the study of materials and their properties comes within the scope of geomorphology is fairly new. Geomorphology has always been difficult to define, but it is unlikely that many geomorphologists would have included the term 'materials' in a definition even 15 years ago. By the mid-1970s, however, one can see the position changing. In 1974, for example, Cooke and Doornkamp were able to write that 'geomorphology . . . is concerned with landform, materials, and their related processes' (1974, p. 7). This is the view that underlies the organisation of the chapters in this book and in this part on pure geomorphology.

As so often, the origins of this new idea go back a long way. Strahler (1952) advocated examining the properties of materials, and Bagnold (1941), Leopold *et al.* (1964) and Paterson (1969) incorporated complex mathematical treatments of the mechanics of particular erosional processes. However, the first important text specifically on the properties of materials was Yatsu (1966), a work which is much better known now than immediately after its publication. The study of materials begins here. The topic was also dealt with, although in a rather different way, by Carson (1971). But the book which brought the subject fully into the mainstream of geomorphological thought was Whalley (1976). This short text is highly original in both content and approach, and is a most instructive read for anyone not already familiar with the topic.

Properties of materials

The essence of the work on properties of materials is this. In geomorphology we are continually confronted with the problem of the circumstances under which erosion takes place and the circumstances under which there is instability in the landscape. This immediately brings us face to face with materials and their properties. It is the way in which rocks and soils behave when stressed which is important. We need to understand the mechanics of that behaviour. Rivers, wind, glaciers, waves and so on merely provide energy; in themselves they are not processes, and just describing them is an insufficient explanation of 'process'. In effect, the study of materials provides the link between orthodox 'process studies' on the one hand and the relationship between rocks and relief on the other. It ties together structural geomorphology and process geomorphology. Traditional structural geomorphology has always considered the properties of rocks in bulk and rather crudely, and, as described above, traditional process studies have always considered the process as the method of erosion, not just as the means of energy input. The study of materials brings the problems of erosion and instability down to the interface between rock and process – to the very surface of materials, at an almost molecular level.

Since we are dealing basically with the point of contact between rock and process, the study of materials can be approached from the point of view of either rocks or processes. It was approached from the point of view of processes by Carson (1971). He was concerned with the mechanics of erosion. He emphasised the similarity in the mechanics of the different sorts of processes which are commonly regarded as diverse topics of study – coastal, fluvial, glacial, aeolian, and so on – and he saw the key problem as being the mechanism by which the loose debris becomes incorporated into the moving fluid or ice mass. He thus unified the mechanics of erosion processes into a single discipline. Scheidegger's (1970) book does much the same, although at a more advanced level. Carson then went on to consider the concept of stress; the relationships between stress, strain and strength; the mechanics of fluvial erosion (in detail, with separate sections on fluid dynamics, the initiation of fluid erosion, rates of fluid erosion, and threshold erosion conditions); the mechanics of glacial erosion, in equal detail; and mass movements in rocks and soils. On the way he made several interesting observations about specific aspects of geomorphology, for example, that the classic problem of the concave long-profile of streams can be explained quite simply in terms of the mechanics of debris movement at a fluid boundary.

Whalley, like Yatsu, approaches the properties of materials from the point of view of rocks and soils. He presents those fundamental concepts in physics, chemistry and soil mechanics which, by explaining the behaviour of materials, can help the student to understand geomorphological processes and landscape formation. After introducing some basic notions on the properties of matter, he considers the bulk properties of materials: the physical states of soils, Atterberg limits, orientation of particles, the packing of particles in a sediment, permeability, porosity, compaction and consolidation. The concepts of stress, strain, strength and shear are discussed, and the basic strain–stress relationships in soils are made clear and put into the context of slope stability. The behaviour of soil particles on a small scale is examined: friction, fracture processes, the properties of clay minerals, and cohesion other than that supplied by clay minerals. He concludes by discussing some geomorphological problems in the light of the behaviour of materials – the strength of ice, rock glaciers, push moraines, frost and other types of rock shatter, and others – stressing that traditional theories are often too simplified. Indeed, we are forced to rethink what we mean by 'explanation' in geomorphology, and it is clear that the properties of materials are an important part of this.

A recent addition to the literature on properties of materials is the work of Selby (1982). In his book on slopes, he brings the methods used by engineering geologists and geomechanics engineers to the attention of geomorphologists. Of particular interest is his chapter on the strength and behaviour of rock and soil. He describes how strength can be measured using both laboratory and field techniques, and considers a number of aspects of rock mass strength: classification parameters; intact strength; weathering; the spacing of partings within a rock mass; the orientation, width, continuity and infill of joints; and the significance of groundwater. The emphasis on the *strength* of surface materials is a distinctive and original feature of Selby's approach.

The appearance of materials as a central theme in geomorphology has implications for other aspects of the subject. These can be listed as follows.

(a) *The relationship between rocks and relief.* The properties of the rocks, as relief formers and in terms of their resistance to weathering and erosion, are investigated in more detail.

(b) *Process studies.* As described earlier, rivers, waves, glaciers and so on are not really processes. They just provide energy. Process is the response of surface materials to these applications of energy, so the crucial point is the properties of those materials.

(c) *Weathering.* It is self-evident that the weathering of a material depends partly on its properties, but it is not self-evident that much remains to be discovered about the role of weathering as a preparatory phase for erosion. For example, studies of stream-channel erosion have tended to focus on the mechanics of debris transport, while relatively little attention has been paid to the preparation of this debris by subfluvial weathering.

(d) *Soils.* In many areas the surface material is a soil, so it is the properties of soils, not just rocks, that are important. This is reflected in Whalley's emphasis on soils.

(e) *Slopes.* The steepness, profile and stability of slopes reflect the properties of the constituent materials.

(f) *Applied geomorphology.* This work on materials is not just of academic interest. It has many practical applications. Three obvious ones are the destruction of natural materials by weathering, ground subsidence, and problems of landslides and other types of slope instability. These and other related topics will be referred to in later chapters.

Conclusion

The study of the properties of materials is a relatively recent entrant into the field of geomorphology. It emerged as a recognisable aspect of the subject with the work of Yatsu (1966), was firmly placed in the mainstream of geomorphology by Whalley (1976), and has been followed up by Selby (1982). It stresses the properties of surface materials in explaining their instability and erosion, and it has a number of important implications for other aspects of the subject.

10 Methods of analysis

Introduction

In this chapter we look at the tools of the geomorphologist's trade. The natural sequence is to begin with the field and laboratory techniques, then to pass on to the range of quantitative, statistical and computer techniques that can be used to analyse the results, and finally to consider the ways in which geomorphologists analyse in a conceptual way the relationships that they study – models and systems. For the most part, the field and laboratory techniques have been developed in response to demand. Necessity has been the mother of invention. On the other hand, some of the numerical and conceptual techniques have been introduced from outside the subject, for example, a number of computer studies. Some of the techniques overlap with other topics in this book. For example, geomorphological mapping, considered in Chapter 14, could be described as a technique.

Although these analytical techniques have been classified in the way described, in practice, of course, the geomorphologist can and does use them in combination. He might use his knowledge of the supposed operation of a geomorphic system to identify a problem that needs to be solved, carry out the necessary fieldwork and laboratory work, and then use some quantitative methods to help him interpret the results. It must also be remembered that all the techniques have their limits and disadvantages. There is, for example, a section in Hanwell and Newson (1973, pp. 8–9) on the importance of designing an experiment correctly, and a long discussion by Pitty (1972, pp. 23–4 and pp. 31–2) just on the difficulties of making a measurement. A final point to bear in mind is that, although this topic has been included in this part on pure geomorphology, all of these methods of analysis are used, perhaps even more importantly, in applied studies (Part IV).

Field techniques

The average geography student at school, with limited experience of fieldwork in geomorphology, would be surprised at the extent to which geomorphology is a field science. There is a recognised technique, or an attempt has been made to find one, for measuring almost every form and process in the field. The best-known textbooks describing these techniques are King (1966) and Hanwell and Newson

(1973), but there are literally hundreds of articles in magazines describing a particular technique, and it is a salutary thought that Goudie's recent tome (1981a) runs to 1750 references! In the face of such a voluminous literature, this discussion can only indicate something of the range of applications.

Field techniques can basically be divided into five categories: those concerned with the measurement of form, properties of materials, processes, sediments and dates. Two of these – properties of materials and the dating of deposits – are, in the main, laboratory studies where fieldwork consists mostly of collection of materials to be taken back to the laboratory for analysis. So we will concentrate on the other three here. All are characterised by techniques varying widely in type and sophistication from the use of bits of orange peel and ping-pong balls as floats, to equipment that can be run up in a woodwork shop in a few hours, to expensive and very accurate devices.

The accurate measurement of form is very basic in geomorphology. Whether it be slope angle or beach profile, the exercise is essentially one of surveying, so the geomorphologist makes use of various pieces of surveying equipment: tape, ranging rod, clinometer, theodolite, and pantometer. At a different scale, measurement of form can include measuring the size of pebbles – pebble morphometry – so calipers are the tool here.

Process studies constitute a massive field, partly because of the variety of processes, and partly because there is often more than one technique for measuring a particular process. This is the case, for example, with the measurement of river velocity and discharge. Almost everyone has experienced the peculiar pleasure of the velocity–area technique, standing thigh-deep in water timing floats over measured distances! (Fig. 10.1.) There is, in fact, a more accurate method based on the same principle, where the velocity is measured

Figure 10.1 Recording present-day processes (from Brunsden & Doornkamp 1977).

with a current-meter. Even this is inaccurate in turbulent water, however, but there are techniques which are accurate only in turbulence – salt dilution, for example. Tracers have a wide variety of uses, including measuring river velocity, longshore drift and the course of underground streams. Pegs or stakes inserted in the ground surface are also of wide application: the movement of soil creep, mudflows and other types of mass movement, and the velocity of glaciers. There are techniques for measuring even the 'lesser-known' processes, such as frost heave (James 1971), and also techniques for measuring processes where there are special difficulties. An example of this would be the use of the Woodhead Sea Bed Drifter (Phillips 1970) to measure sea-floor water movement. The measurement of overall rates of erosion has thrown up a whole range of techniques, including study of old maps, pegs embedded in the ground, and drawing inferences from the sediment content of rivers (High & Hanna 1970).

Work on soils and sediments is a whole area of study in itself. Measurements on soils include pH values, soil moisture (Knapp 1973) and infiltration capacity (Hills 1970). Of several techniques pertinent to glacial and fluvioglacial deposits, till-fabric analysis (Andrews 1971) is perhaps the best-known. The measurement of the load of rivers poses special problems, and there are separate techniques for the calculation of solution load (Douglas 1968), bed-load (Gregory & Walling 1973, p. 164) and suspended sediment (ibid., pp. 152–4). In work on sediments there is a close link between fieldwork and laboratory work, the fieldwork consisting of augering, and the laboratory work the analysis of the samples in the core.

Remote sensing

This is also a field technique, but it is in a completely different category to those discussed in the previous section in that it is highly specialised and does not form part of the normal hardware at the disposal of geography or geology departments. There was considerable activity in this area by the mid-1960s – see, for example, Latham (1966) and Simpson (1966) – and the rapid pace of development since then is shown by the appearance of whole textbooks on this one subject, for example, Lillesand and Kiefer (1979) and Townshend (1981). The technique is basically one of remote sensing of the environment by sensors carried by aircraft or satellites, followed by image analysis and interpretation by both manual and automated methods. The application of remote sensing is increasing all the time, but is especially valuable in the initial reconnaissance stage of terrain analysis and evaluation with a view to assessing the resources, relief, structure, surface materials, vegetation cover, hydrology and even subsurface properties of an area. Several of these topics are considered in the later chapters on applied geomorphology. Remote sensing would seem to have its greatest value in hitherto 'uncharted' terrain where orthodox ground fieldwork is difficult, but interestingly Townshend (1981) maintains that a paucity of ground data has hindered the expansion of remote sensing. The two obviously go together.

Laboratory work

In many areas of research work, geomorphology is increasingly becoming a laboratory science, although there is no prospect of it becoming almost exclusively so like physics or chemistry. There are basically two types of laboratory work in geomorphology.

The first is where an experiment is being carried out under laboratory conditions, with a view to the results being of wider geomorphological significance. A well known example of this is the experiment of Griggs (1936) demonstrating the importance of moisture in isolation weathering, and a more recent one the experiments by Goudie *et al.* (1970) to investigate rock weathering by salts.

The second type of laboratory work is where fieldwork needs to be followed up by some kind of laboratory analysis. One example of this is where a series of sieves is used to sort a sediment sample into its grain sizes. Another is where titration is used to calculate the hardness of a sample of limestone water. Sometimes the laboratory work aims to solve a particular problem and requires expensive apparatus. For example, one technique for establishing the origin of a deposit (i.e. whether fluvial, glacial, fluvioglacial, marine or aeolian) is to subject the surface of the individual grains to minute examination under the microscope.

However, probably the main application of laboratory work is in the analysis of Pleistocene sediments. This includes correlation with other samples, dating, fossil analysis, and the identification of the environment in which the deposit accumulated. Our knowledge of the sequence of events during the Pleistocene (Ch. 6) rests largely on this sort of laboratory study. The work involves not just the well known radiocarbon dating, but a whole host of other techniques such as dendrochronology, the analysis of varves and other sediment cores, tephrochronology, pollen analysis, and work on fossil beetles and non-marine mollusca.

Quantitative geomorphology

The progressive introduction of quantitative techniques into geomorphology since about 1945 has been one of the most evocative and controversial recent developments in the subject. Quite understandably, older and more traditional workers, especially some schoolteachers, brought up on qualitative and Davisian principles and with limited mathematical backgrounds themselves, felt annoyed, puzzled and even threatened by the quantitative 'revolution'. Even some university academics were hostile or at best showed a sceptical interest, and were concerned that the approach might lead to the dogged analysis of trivia and to a decline in true geomorphological scholarship. Even today there is a less than 100 per cent commitment from the academic community, and there are two reservations with which few would disagree: there is a danger of the elaborate and complicated 'discovery' and re-statement of the well-known, and the techniques must always be seen as a means of analysing a geomorphological problem and must not become the main object of the exercise in themselves.

A number of different terms have been introduced as a name for the topic being discussed here: one can read of quantification, numerical analysis and statistical techniques. This is not surprising since the application of numbers to geomorphology has taken place at several different intellectual levels. At one extreme one can measure a slope and say it has an angle of 60° instead of merely describing it as steep, and at the other one can program a computer to perform a spectral analysis of a meander train. But, in so far as we can identify an area called 'quantitative geomorphology', it all comes under the same umbrella. Some levels of analysis yield, or make use of, either equations or laws. This can be a touchy subject. It is sometimes said that the very complexity of geomorphic systems means that the precision of mathematical relationships does not exist everywhere in geomorphology. This is true, and the implications of it are, firstly, that equations are inappropriate in such circumstances and, secondly, that in geomorphology we use the term 'law' rather loosely by scientific standards. For example, Horton's well known 'law' of stream numbers is not a law at all in the sense that Newton's Law of Gravity is a law. But, on the other hand, some geomorphological relationships can be written quite legitimately as equations, in which case their use – as a mathematical shorthand, as in other sciences – is totally justified.

There was no single reason for the introduction of quantification, nor even one main one. Rather, there was a series of other developments in geomorphology with which numerical analysis went hand in hand: opposition to Davisian thinking, the appearance of process studies and applied studies, the introduction of field and laboratory techniques that required statistical analysis, and the increasing links with other disciplines, such as hydraulics, where the use of mathematics was in existence already. It was also seen that the only sure way in which two workers in different parts of the world could compare their observations of form or process was to measure those observations accurately, and that if geomorphology was to become a predictive science then the required precision could come only from quantification. There were also similar developments taking place in human geography as well, so that the new moves in geomorphology merely reflected what was happening in geography in general. And, finally, as so often in science, the successful adoption of the new approach was to some extent a personal success for workers who had vigorously promoted it: in this case, Strahler in America (see, for example, his 1954 article) and Chorley (1966) in Britain can lay claim to such fame.

It is assumed that virtually all readers of this book will be familiar, at the very least, with the statistical measures covered in A-level geography syllabuses, and that most will know of the basic textbooks, at various levels, on quantification in geomorphology: Gregory (1963), Cole and King (1968), Doornkamp and King (1971) and Hammond and McCullagh (1978). In view of the existence of these books, no attempt will be made here to describe all the techniques that can be applied. Instead we will trace the diffusion of statistical concepts into geomorphology.

We have now reached a stage where quantitative techniques are more or less essential to underpin other aspects of geomorphology: morphometry, process studies, applied studies, and a whole range of field and laboratory techniques. They are taken for granted. But the assimilation of statistical concepts into the subject was not a smooth process: instead it was irregular and staccato.

It is difficult to say when quantification began, because many early studies used numbers in one way or another. To many geomorphologists, however, the classic work of Horton (1945) marks the beginning of the new era, while the pioneer in the application of statistical techniques was Strahler (1950). It seems safe to refer to quantification as a post-war phenomenon.

During the 1950s geomorphologists began to use a wide range of 'linear' techniques (as opposed to spatial techniques), especially correlation and regression analyses and analysis of variance (Chorley 1966). Towards the end of the 1950s more complicated types of multiple regression began to be adopted, at first by means of laborious manual calculation and then in association with second-generation computers. This meant that by about 1960 geomorphology was technically ready for a number of advances which would bring it more into line with human geography and geology, which had developed rather earlier. These advances could be in the fields of computer-based quantitative

113

techniques, of models and especially of probabilistic models, and of spatial analysis.

It is at this point, however, that something rather curious happened. It is described in detail by Chorley (1972, Ch. 1). A number of these advanced quantitative techniques were applied to geomorphological problems by individual innovators in the subject by the mid-1960s, but then there was a delay until the early 1970s before the techniques passed into more general use. Examples of these advanced techniques are sequential multiple regression, polynominal trend-surface analysis, harmonic and spectral analysis, factor analysis, Markov-chain analysis, discriminant analysis and principal components analysis. They all had their 'première' in geomorphological literature sometime between 1960 and 1965.

Then came the delay, however. It was a delay both in applying these techniques to established problems and in formulating new problems appropriate to the new analytical tools. The reasons for it are given by Chorley (1972, pp.6–8).

The take-off year in which the delay could be seen to be over and the advanced techniques were beginning to be applied more freely has been identified by Chorley as 1971 (1972, p.3). There is some justification for this. Chorley and Kennedy (1971) described the range of statistical concepts that can be applied in the analysis of the various geomorphic systems, and Chorley (1972) demonstrated how widely techniques of spatial analysis were being used in the subject by that time.

A recent review of quantitative geomorphology by Morisawa (1981) confirms that we stand in a position where a very wide range of numerical techniques both simple and advanced, are used by geomorphologists. No doubt there will be new developments in the near future, but it will be interesting to see what happens to the current assemblage of techniques. It often happens in geography that, over a period of years, new ideas gradually filter down from innovator via research workers and undergraduate courses to schools. If this happens here, the schools should be able to see what is coming.

Models

When Chorley and Haggett produced *Models in geography* (1967a) it created enormous interest among geographers. It was not that models were new; indeed, the vast bibliography showed that quite the reverse was the case. But all the relevant work had never been brought together in that way before, nor had models been seen as a unifying element in geography, and neither had models been viewed as a framework for future research. The model-based approach, in con-

114

junction with the adoption of various quantitative techniques, prompted Manley (1966) to speak of the 'New Geography', a term which passed into popular usage, although coming to mean all things to all men.

A model is a simplified representation of reality made so that we can understand the environment more readily. It presents only the supposedly significant features or relationships, and those in a generalised form. Incidental detail is eliminated so that the fundamentals stand out. It is an analogy, a simulation of reality. A model can take a wide variety of forms. It can be a theory, law, hypothesis, relation, an equation or a synthesis of data. The purpose of model-building is clear enough: the extreme complexity of reality means that some measure of idealisation and simplification is a welcome aid to understanding.

Chorley and Haggett showed that the model-based approach was appropriate not only to geomorphology (Chorley 1967), but to related subjects such as hydrology (More 1967), other aspects of physical geography, to geography in general, and even to science as a whole (George 1967). One effect of all this, therefore, was to strengthen the links between geomorphology and the rest of geography, and to stress the scientific aspects of geography.

There are many different types of model, and almost every textbook classifies them in a different way. However, from the point of view of the application of models to geomorphology, a four-fold classification is appropriate.

(a) *Natural analogue models*. The supposedly important or characteristic features of a geomorphological phenomenon are translated into some analogous natural system believed to be simpler, better known or in some respect more readily observable than the original. Large sections of geomorphic work traditionally lie in this area, and two examples are the cycle of erosion and denudation chronology, neither of which are normally thought of as models. Unfortunately they involve so many built-in assumptions that any testing to which they have been subjected usually develops into circular reasoning (Chorley 1965).

(b) *Hardware models*. This is the sense in which even the layman would understand our use of the term 'model': the physical construction of a scale model. They fall into two categories: iconic models, where the materials used in making the model are the same as the materials in the real world, and there is a change only in scale; and analogue models, where the material properties are changed as well. An example of the former would be if sand and water were used to model a beach. There are considerable problems, not least because, although the model is scaled down, the particles of sand and the water are not. An example of an analogue hardware model is Lewis and Miller's (1955) attempt to model glacier behaviour using kaolin. Another example, more recent, and designed to solve a particular problem, is the scale coastal model constructed by Hydraulics Research at Wallingford in order to study the silting of Teignmouth harbour, Devon. It is discussed by Hanwell and Newson (1973, pp. 174–6). Even with this type of model there are difficulties with obtaining results that can be applied in some useful way (see, for example, Morgan 1967), and this probably explains why the technique had not been explored more fully by the mid-1960s. They are certainly of wider application now, mainly because of their usefulness in simulating engineering problems. Some examples are given in Chapter 13 of the use of hardware models in applied geomorphology, and Cooke and Doornkamp (1974) discuss even more.

(c) *Conceptual models*. Here, the model is not a physical object at all, but is some sort of mental abstraction. Most of the models in this category can be described as mathematical models, and there are two types. The first is the deterministic model, which is based on a precise mathematical relationship existing between two sets of factors. The model could, therefore, be an equation or a law, and the predictive value of such a model can be considerable. An example well known to most readers would be the expression of the long-profiles of streams in the form of some sort of logarithmic relationship. The second type of mathematical model is the stochastic model, where the modelling procedure involves a random element. An example of this would be Leopold and Langbein's (1962) attempt to reproduce a drainage basin network using a random walk (Fig. 10.2).

(d) *Systems-based models*. Geomorphologists express geomorphic systems as models. For example, Figure 10.3 shows the major factors involved in the hydraulic geometry system. The diagram is a model. There is, therefore, a close link between models and systems.

To be of any practical value models must be testable. The model is tested against the reality that it simulates, and analysis of the disparity between the two means that the model can be modified to give a closer approximation to reality. There is a limit to this process, of course; modification usually means complication, so as reality is approached the essential simplicity of the model is lost. However, such re-modelling remains an important part of model-building, and in any case the existence of the model means that in studying reality, one knows what one is looking for.

116

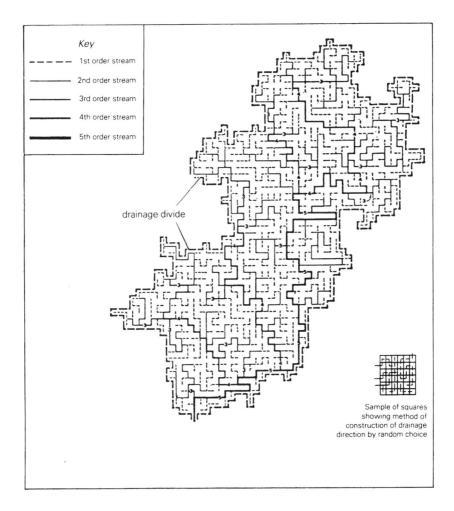

Key
- - - - - 1st order stream
———— 2nd order stream
———— 3rd order stream
———— 4th order stream
━━━━ 5th order stream

drainage divide

Sample of squares
showing method of
construction of drainage
direction by random choice

Figure 10.2 The development of a fifth-order drainage basin network by a random-walk method from a series of squares, each of which has an equal chance of draining in any of the four cardinal directions (from Leopold & Langbein 1962).

It must be admitted that there are pitfalls in modelling. This was foreseen by Chorley and Haggett themselves in 1967. They write (1967, p. 26):

Simplification might lead to 'throwing the baby out with the bathwater'; structuring to spurious correlation; suggestiveness to improper prediction; approximation to unreality; and analogy to unjustifiable leaps into different domains . . . a bad model would be heavily symbolic, present an over-formalised view of reality, be

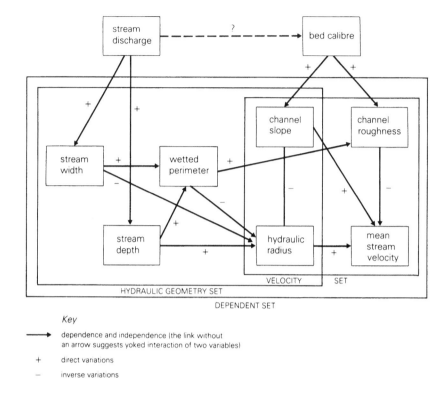

Figure 10.3 General relationships between the major factors involved in hydraulic geometry (from Chorley & Haggett 1967).

much over-simplified, represent an attempt to erect a more exact structure than the data allows, and be used for inappropriate prediction.

It is also true that, since models often present a very attractive simplification of reality, they might be substituted for and accepted as reality. One must also bear in mind that, since reality can be simplified in many ways, one particular model is just one possible way among many of representing reality.

A rather more fundamental and philosophical criticism of the model-building approach is the view, long held in geography, that the complexity of reality means that we should study our subject as a number of unique situations rather than in terms of generalisations. It is reservations of this sort that have prompted the current dissatisfaction among some university geographers with the model-building approach. In some circles, model-building is out of favour, and replacement philosophies are argued by structuralists, Marxists and humanists. This is not a discussion to get drawn into here. The attack, such as it is, is on geographical rather than geomorphological models, and it would seem that model-building, within the systems framework, continues to offer profitable returns in geomorphology.

Systems analysis

When Chorley and Kennedy's *Physical geography: a systems approach* appeared in 1971, systems analysis was not new. Its origins go back to the work of von Bertalanffy (1956). The pioneers in its application to geomorphology were, in America, Strahler, and in Britain, Chorley (1962), the same workers who initiated the model-based approach. Chorley and Kennedy's (1971) vast bibliography showed how much work had been done along these lines by the early 1970s. The effects of the book were to draw all this work together, to bring it to the attention of academics in general, and to promote systems analysis as a profitable way of organising geomorphological information. Since then, a host of textbooks and other work have used the systems approach as a basic methodology, for example, Bennett and Chorley (1978).

The meaning of the term 'system' becomes more obvious by example than by definition. However, we can define it as any structured set of objects and/or attributes, together with the relationships between them. This rather abstract definition tends to make the whole subject of systems analysis seem rather obscure. But this is not so, especially if one remembers that the systems approach is not in itself geomorphological data but a method of analysing that data.

There is more than one way of classifying systems. A common one is a threefold classification into isolated systems (where the system is closed to the import and export of both mass and energy), closed systems (which are closed to the transfer of mass but not energy) and open systems, where both mass and energy are exchanged. Most natural geomorphological systems are open.

Geomorphological systems can be classified into four types on the basis of their internal complexity. In increasing order of complexity, they are as follows.

(a) *Morphological systems.* These consist purely of structural relationships between the constituent parts of systems. An example of this is where first-order, second-order, third-order (and so on) streams combine to form a drainage system. This particular example is a network, and it is through networks that most people are familiar with the system concept because of our use of terms such as 'transport system'. However, not all morphological systems are networks in the sense that they can be represented on a map like a drainage system. Three other examples are slope morphometry, hydraulic geometry and beach morphometry. Each system is identified on the basis of there being significant correlations between different elements in the system.

(b) *Cascading systems*. These systems are defined by the path followed by throughputs of energy and mass. An example is the stream-channel cascade, where water, debris and solutes cascade down through the system. Other examples are the weathering cascade, vegetation cascades, debris cascades, valley-glacier cascades and wave cascades. In studying such systems, emphasis is placed on the relationship between the input at the head of the cascade and the output at the end of it. There are three ways of analysing such cascades, and they will be illustrated with reference to the hydrological cascade:

   (i)   *Black box*. The input (precipitation) is related to the output (streamflow) without there being any attempt to analyse the factors linking the two.

   (ii) *White box*. An attempt is made to identify all the controlling factors operating within the system – various flows, storages and so on; in this example, this is the terrestrial part of the hydrological cycle (Fig. 10.4).

   (iii) *Grey box*. A partial analysis of the system itself is attempted.

(c) *Process–response systems*. Technically this is the intersection of at least one cascading system with at least one morphological system, but most geomorphologists will comprehend it instantly if we look at it as any system whereby a particular process produces a particular response (form). This is process-form studies in systems terms.

(d) *Control systems*. This is a process–response system in which man has intervened in some way. Therefore when, in later chapters, we look at man's intentional and unintentional impacts on landscape, we are in effect studying control systems.

An understanding of the operation of systems is, in some ways, more important and more interesting than knowledge about the different types. A significant property of open systems is that the elements within them tend to adjust themselves towards some sort of equilibrium condition. The term 'steady state' is reserved specifically for the situation where the input of mass and energy balances the output. Equilibrium does not mean that the system is static, however.

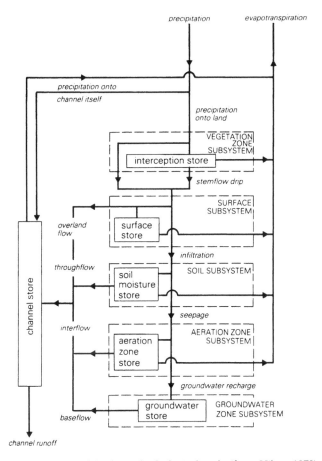

Figure 10.4 Components of the basin hydrological cycle (from Hilton 1979).

It is performing work all the time, and opposing forces are continually fluctuating about a mean position – dynamic equilibrium.

A fundamental mechanism in maintaining this state of self-regulation is feedback. This means that when one of the components in a system changes, there is a sequence of changes in the other components which eventually affects the first one again. The most common type of relationship, and the one underlying equilibrium in open systems, is negative feedback. The circuit of changes has the effect of damping down the first change. Most students will be familiar with this idea via the concept of the graded river (Mackin 1948). Positive feedback, which is rarer, accentuates the original change. It tends to operate in short bursts of destructive activity. In the long term, however, negative feedback and self-regulation tend to prevail.

At this point it needs to be made clear that the concept of

equilibrium in systems has prompted research and intellectual thought on a number of related topics. One is that a common method of recognising the existence of some kind of equilibrium is the identification of significant statistical correlations between variables within systems. Another is the idea that natural processes constantly seek economy of effort or maximum efficiency. River meanders and long-profiles have been analysed in terms of entropy (Leopold & Langbein 1962) and of energy dissipation (Langbein & Leopold 1964).

However, not all systems are in equilibrium. The main reason for this is that there is a delay between change of input and response in the system. This delay is called the 'relaxation time', and interestingly it varies enormously from one system to another. For example, a beach profile might change because of a storm in a matter of hours. The short relaxation time is, in effect, why beaches are so suitable for studies of present-day processes. In contrast, the relaxation time for glacial features such as corries in Britain in response to present-day processes is so great that the landforms are virtually unmodified after 10 000 years. That is why form is out of phase with process.

There is no doubt that the systems approach is currently proving to be a very suitable framework for a wide range of geomorphological work. As discussed in the final chapter in this book, it might even be fulfilling the role of a paradigm in the subject. Whatever the truth of that, it must also be conceded that systems analysis has not been adopted wholeheartedly by everyone. The reasons for that are probably as follows.

(a) Natural resistance to anything new by some older, more conservative, workers.
(b) Difficulties with understanding systems analysis. The main textbooks are aimed at the highest intellectual level and make no attempt to offer a plain man's guide. There is also some difficulty with some of the terms used, such as filters, regulators and valves.
(c) The association of systems analysis with the statistical procedures used to analyse them.
(d) Systems analysis is not central to all geomorphology. For example, it plays little part in studies of Pleistocene chronology.

There is one danger that everyone must beware of – the danger that we might mistake the framework for the reality, and set off trying to identify systems rather than use the concept as an aid to understanding.

Thresholds

There is one concept in systems analysis which, during the last 10 years, has been developed to the point where it now occupies a central position in geomorphological thought in its own right. This is the 'threshold'. The concept was referred to repeatedly by Chorley and Kennedy (1971), the threshold terminology in geomorphology was originated by Schumm (1973), and the wide range of applications of the idea were brought together by Coates and Vitek (1980). The threshold has long been an important concept in disciplines such as metallurgy, physics and chemistry, so the new work not only puts the threshold concept into the mainstream of geomorphological thought but also puts geomorphology into closer contact with other sciences.

A threshold is a condition in a system marking the transition from one state or economy of operation to another. It can be a critical limit, a boundary condition, a yield point. The system 'flips' from one state of operation to another. Some examples will help to make the idea clear. Examples can be drawn from almost any branch of geomorphology, illustrating that the threshold can form a conceptual base for virtually the whole subject.

The most obvious examples of thresholds concern hazards. For instance, an increasing discharge fills a river channel to the point where bankfull discharge is exceeded, and overbank flow occurs. The threshold is bankfull discharge, and the threshold-exceeding event is a river flood, which is a hazard. Another example occurs where an increase in runoff is sufficient to strip off the surface vegetation and initiate a system of gullies which cannot be obliterated even after a prolonged return to more normal runoff conditions. A third example concerns slope stability. All types of slumping, landsliding and so forth can be considered as the response of a slope system when a critical threshold is exceeded: the forces of disequilibrium suddenly overcome the forces of equilibrium. The system changes from one equilibrium situation to another. The threshold is a basic concept to all problems of properties of materials and stability. A change occurs when the threshold value of stress or strength is exceeded. A critical slope angle can often be identified as a threshold: for instance, it has been observed that slips in the London Clay occur only when the slope angle exceeds the critical threshold value of 10°.

The threshold concept can also be used in a rather different sense. Some very familiar geomorphic situations can be seen as threshold-exceeding events once one gets used to the idea, for example, river capture and the formation of meander cutoffs. Sometimes the threshold is evident only when data are presented in a certain way. For example, the fairly well known graph in Figure 10.5 shows that a

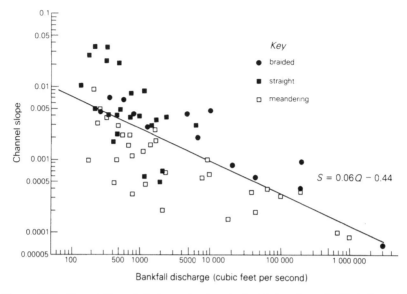

Figure 10.5 The relation of discharge to slope in braided and non-braided rivers (after Leopold & Wolman and Balek & Kolar: from Leopold *et al.* 1964).

channel slope: bankfull discharge threshold separates braided river channels from meandering ones. Sometimes the existence of the threshold is inferential. One example is that natural streams usually have channels of just three types: bedrock-floored channels, fine-bed alluvial channels and coarse-bed alluvial channels. Very few stream-reaches exhibit any sort of intermediate characteristics. The inference is that the change between the channel types is attributable to the behaviour of sediment transport and deposition. A second example is that the tendency for granites to form either mountains or plains in the tropics is possibly governed by a threshold value in the balance between the rate of surface erosion and the rate of subsurface chemical weathering.

One thing that emerges from all these examples is that in most, but not all, situations, the passage across the threshold is irreversible. The landslide does not afterwards go back to its previous position. This is interesting when one looks at systems in feedback terms. The self-regulation of negative feedbacks is suddenly broken, a threshold is exceeded, then the system adjusts to a new and totally different dynamic equilibrium.

The threshold concept is of both intellectual interest and applied value. There are, in fact, two types of threshold. The first is intrinsic. The threshold is crossed due to influences internal to the landscape. The river floods because of heavy rain. The second is

124

extrinsic. The influence is external. Man is a possible external influence. He can cause thresholds to be exceeded. He can also sometimes prevent that from happening. Therefore, one way of looking at environmental management is that man is intervening to prevent thresholds from being exceeded. With these thoughts in mind we can pass naturally to the next major field in geomorphology – applied studies.

Conclusion

Part III of this book, on pure geomorphology, has been brought to a conclusion with a review of the vast range of techniques available to the geomorphologist. These include fieldwork, remote sensing of the environment and laboratory work. A wide definition of the term 'technique' has been adopted so as to include the numerical techniques used in the statistical analysis of data, and also the conceptual techniques used as a framework for the subject.

PART IV

Modern applied geomorphology

11 Background to applied studies

Introduction

Contrary to popular belief, applied geomorphology has a long history. In Chapter 1 we saw that the origins of geomorphology were marked by numerous applied studies between the 17th and 19th centuries, not least those by Gilbert himself. This tradition was carried on into the first half of this century by, for example, Glenn (1911), Sherlock (1922), Bryan (1925), Horton (1924, 1932, 1933, 1945), and Jacks and Whyte (1939). Thornbury (1954) also devoted an excellent chapter to applied geomorphology in his widely read textbook. In spite of these works, and others, it is clear that even as recently as the 1960s geomorphology did not have a strong applied content. For example, Dixey (1962) was prompted to stress the need for more geomorphological studies to be orientated towards the needs of man, and Dury (1969) conceded that geomorphology did not include numerous applied contributions.

It is not unusual in science for an idea or an approach, or even a whole branch of a subject, to have early origins but a late development. Even so, we must ask why it happened with applied geomorphology. No doubt there are several reasons. Two that come readily to mind are the long association of geomorphology with geography at a time when geography itself had little applied content; and the long preoccupation with the cycle of erosion and denudation chronology, neither of which had any obvious applications to practical problems. But probably the most important reason was the delay in the development of process geomorphology. To a large extent, applied studies and process studies go together, because applied studies rely on there being an explanation for geomorphological phenomena, and explanation is impossible without understanding process. Now, as we saw in Chapter 8, there were important process studies in the first half of this century – in fact, modern process geomorphology goes back to Gilbert (1877) – but in general the results were not applied to practical problems even though they could have been, and the main drift of geomorphology was in another direction anyway. So applied geomorphology had to wait for process geomorphology, and that meant waiting until the 1960s, although there were attempts to develop process studies in the 1950s. In round figures, we can say that by the end of the 1960s process geomorphology was strong enough for applied geomorphology to take off. By then, other relevant developments had also taken place: landform and landscape evaluation, the land systems approach and geomorphological mapping.

Around 1970 there was a frenzy of activity, showing that geomorphologists had realised that the results of process studies were of practical as well as purely academic value, and showing that applied geomorphology was being born as a recognisable discipline: in 1967, Douglas's article on man's effect on the sediment yield of rivers; in 1969, Chorley's widely read *Water, earth and man* (1969a); in 1970, Brown's much-quoted article on man as a geomorphic agent, and Craik's pioneer work on man's perception of the physical environment, indicating that the relevance of geomorphology has to be seen in the context of man's response to what he *thinks* the environment is like rather than what it is *actually* like; and in 1971, Detwyler's *Man's impact on environment*, and Douglas's two widely quoted case-studies; and Chorley and Kennedy's *Physical geography: a systems approach*, not only a milestone in itself, but showing that man's impact on landscape has to be seen in the context of his place in complex systems; and Clayton's well known article stressing how important it is for man to understand the environment well enough so that he can predict the implications of his interference with it, stressing that the physical system of an area can never be understood fully in isolation from the social, cultural and economic attitudes and conditions of the people of that area.

Then, as a culmination of this activity, came the publication in 1974 of Cooke and Doornkamp's *Geomorphology in environmental management*, a masterly work defining a coherent area of study, summarising a vast existing literature, and pointing the way forwards. Applied geomorphology is now the main area of growth in the subject; over the past 10 years or so there have appeared countless magazine articles and textbooks, or chapters in textbooks, for example, Leggett (1973), Tank (1973), Hails (1977) and Coates (1981).

The need for applied studies

Since geomorphology has not always had a strong applied content, but instead has been mainly an academic discipline, it is pertinent to ask why there is a need for the applied branch of the subject. What justification is there for it, and what demand is there for it? The answer is that there is a need for applied geomorphology from two quarters. One is an intellectual demand from within geomorphology itself; the other is a practical demand from society in general.

The first is a general feeling among geomorphologists – perhaps one should say a recognition by geomorphologists – that the subject cannot survive as a purely academic discipline. The subject must be justified, at least in part, in terms of its practical value to the community. There are three reasons for this. First, government money is being spent on undergraduate courses and postgraduate research, so there should be some tangible return from these courses and researches. Secondly, it would be absurdly wasteful to have a subject which could make practical contributions yet did not do so. Imagine the situation if medicine, for example, isolated itself as a purely academic subject and failed to make any of its research available for public use! Thirdly, a number of geomorphologists have seen applied studies as the ideal outlet for their expertise and have deliberately set about promoting the subject. Applied geomorphology did not just 'happen'.

However, it might have just 'happened' anyway, because the second reason for the development of applied geomorphology has been a genuine demand for it from outside the subject, from the community in general. This demand takes a number of forms, and this in turn defines what applied geomorphology consists of, there is no point in providing a service that nobody wants. It is, therefore, possible to classify modern applied geomorphology into various branches and, although the branches have links and overlaps, they provide a convenient basis for describing the subject.

The first branch is concerned with natural hazards. These include soil erosion, various types of slope failure, sea and river floods, volcanoes, earthquakes and faulting. Sometimes the hazard occurs, at least partly, because man himself has acted unwisely in some way. This is often the case, for example, with soil erosion. Sometimes the hazard is, or appears to be, completely natural – earthquakes, for example. Whatever man's own role in the initiation of the hazards, however, their combined effects are enormous: apart from the loss of life, incalculable sums of money are involved in the damage caused and in the insurance and legal claims that follow (Coates 1980). The professional geomorphologist has a role to play here. He has some measure of understanding of the combinations of events that produce the hazards, and he is therefore in a unique position to advise on such matters as predicting the occurrence of a hazard, protecting against it so as to reduce its effects, and perhaps in reducing the dimensions of the hazard itself.

The second branch is environmental management. This is the geomorphologist's role in man's deliberate and controlled impact on the landscape: the geomorphologist joins others in planning, managing and developing the environment. There are close links with hazard control, and in a sense environmental management includes all the work on hazards, but environmental management does not necessarily imply the presence of a major hazard. For example, geomorphologists are called upon to advise on the siting and building of roads and settlements, and this may or may not involve a potential hazard. In this area, man is trying to 'manage' the environment in ways that are beneficial to himself whilst minimising further problems that might result from his intervention.

The third branch is the evaluation of resources. There is a general concern that we should be aware of the existence of the resources that are available to us, and that those resources should be the subject of suitable conservation measures. The geomorphologist has a variety of roles here. First, sometimes a resource has a particular geomorphological setting, as with the occurrence of sand and gravel deposits as river terraces, so that the evaluation of such deposits has an obvious geomorphological element. Secondly, at a more general level, most land resources are closely linked to the geomorphology of the Earth's crust, so the trained geomorphologist can assess rapidly the resource potential, including mineral resources, of an area from the interpretation of its landforms. This is particularly valuable in previously uncharted terrain. Thirdly, the geomorphologist is also a conservationist. Conservation is important for all resources, but especially perhaps the soil, and the geomorphologist, with his knowledge of what creates and destroys soil, has a crucial role to play here. Conservation is also important for some things that we do not immediately think of as resources – scenery, for example. Therefore, the techniques of scenic evaluation are part of environmental management. Fourthly, terrain analysis is also important for some things that we do not immediately think of as geomorphological. For example, it has military applications. Not only is an appreciation of terrain instilled into every soldier during his basic training, but at a higher level classifications of landscape now play an important part in the planning of military operations, as in the Falklands War of 1982.

The demand for geomorphological information comes from a variety of organisations concerned with developing or managing the environment. These may be government bodies, including authorities, or consulting engineers, land developers, land managers, and those involved in the legal aspects of planning.

The type of information that the geomorphologist can, and is, asked to supply depends partly on the subject of the investigation but also on the scale of the planning or development problem. At a national level the emphasis tends to be on resource inventory and appraisal, leading to the selection of suitable regions and locations for development. At the regional and local levels more importance is attached to detailed field mapping of geomorphological features and to process studies as bases for determining the risks associated with development. More detailed investigations are required at the scale of the individual site when the proposals of the plan are implemented. Studies at this stage are concerned with the nature of the surface materials, slope stability and earth-moving or land-forming operations. As a planner, therefore, the geomorphologist is making a variety of contributions at different times and at different scales to numerous organisations and authorities. This is illustrated in Table 11.1.

Whether the geomorphologist is acting as a private consultant or a full-time employee, he brings a unique combination of skills to the job. They arise from his basic training as a geographer, as is usually the case in Britain. They have been listed by Brunsden *et al.* (1978) as:

(a) An ability to think in spatial terms: to appreciate location, and to cope with several phenomena at a time.
(b) An ability to detect spatial correlations.
(c) An ability to change one's scale of thinking in accordance with the nature of the problem.
(d) An ability to comprehend the significance of the time dimension.
(e) An ability to use plan documents: the map, plan or aerial photograph.

It might well be possible to add to this list, since geographers also have an eye for country, an ability to synthesise, an appreciation of the role of man, and other skills as well. These are the advantages that the geomorphologist has over other scientists such as engineers. They are the reasons why the geomorphologist is in demand.

Relevant contemporary developments

Over the past 10 years or so, applied geomorphology has gradually gathered momentum to the point where it is now a major branch of the subject. But it has not developed in isolation. There have been other developments, some within the subject and some outside it, which have taken place just before it or at the same time. We cannot pinpoint any one of these other developments and say that alone it caused the appearance and growth of applied geomorphology, but taken together they allowed it and stimulated it. The link with applied geomorphology is obvious and strong, but it is difficult to speak of cause and effect.

Of the developments within geomorphology itself, the first, the rise of process studies, has been discussed already. The second is closely related to it: the development of numerical techniques and their application to the subject. This has been described in Chapter 10. The relevant aspect of this development for applied geomorphology is that

Table 11.1 The professional roles of the applied geomorphologist.

Nature of employment	Consultant to an organisation, or full-time employee of an organisation. Importance of communicating his knowledge to the public, pressure groups, various agencies, government departments.		assessing dimensions of potential hazards. Land grading and land-forming requirements.
		Construction stage	Implementing the plan: advice given on nature of surface materials, problems of slope instability, earth-moving and land-forming operations.
Reconnaissance stage	Exploratory and reconnaissance surveys, possibly at a national level. Typical mapping scale would be 1 : 250 000. Might involve land systems mapping. Resource inventories and appraisals. Techniques of scenic evaluation: landscape as a resource; assessment of conservation issues. Project location: selection of suitable regions and locations for development.		
		Predictions	Prediction of the effects of plan implementation, of man's impact, of environmental impact, e.g. increased erosion rates. Preparation of environmental impact statements. The geomorphologist can think in terms of side-effects and wider consequences.
Feasibility stage	Project feasibility and planning. Land classification and evaluation. Geomorphological mapping. A typical scale would be 1 : 25 000 to 1 : 50 000. Location of trial pits and boreholes. Feasibility studies completed before the decision to invest is taken.	Monitoring	Monitoring the effects on the environment of plan implementation, the dynamic changes in the environment. Advising on any new measures needed.
		Professional advice on related issues	Formulation of legislative or other controls over planning. Drawing up site ordinances and building codes. Legal questions: the geomorphologist could appear as an expert witness in the courtroom; case might involve problems of erosion or sedimentation, damage to property and water supply caused by construction, putting a price on superficial deposits.
Site investigation stage	Detailed development surveys and site evaluation. Land suitability. Process studies: application of knowledge from pure research, erosion plots, experimental stations etc. Hazards: mapping of existing hazards;		

it allows the geomorphologist to speak in precise, numerical terms all the time. The applied geomorphologist has to deal in such precision. When there is a problem to be solved, and people's lives or vast sums of money are involved, then it is not enough to speak only in qualitative terms.

The third development is the increasing awareness of man as a geomorphological agent. This also is nothing new, but an appreciation of the sheer scale of man's intervention is new. Man's intervention has often been ill judged and harmful. Man himself therefore seeks to take corrective measures and to avoid making such mistakes again. And already one is talking about environmental management, at the very core of applied geomorphology.

The fourth development is the emergence of general systems analysis as a paradigm for geomorphology. General systems analysis provides an accurate and logical framework for appreciating and studying the complexity of the geomorphological environment. One is virtually never dealing with simple cause and effect. Nearly always there is a variety of factors involved, factors which interlink and affect each other. So if a practical problem needs to be solved, this complexity needs to be not just recognised but accurately understood. General systems thinking allows this.

A fifth and final internal development has been the appearance of an abundance of techniques. The applied geomorphologist, perhaps faced with a landform or process to measure that he has not measured before in an environment that he has not worked in before to solve a problem that he has not thought about before, needs a veritable armoury of techniques at his disposal, and this is exactly what he has got (Ch. 10). The proliferation of techniques does reflect the demand for them, of course. Some of the techniques have been developed by geomorphologists themselves, while others are the products of science in general. It is perhaps inappropriate to single out one example when there are so many, but it illustrates the general point to observe that the applied geomorphologist's job of resource evaluation is greatly facilitated by the development of the techniques of remote sensing.

The second group of developments that need to be stressed are those that have occurred outside geomorphology but have still gone hand in hand with the growth of applied studies. The first of this group would be regarded by some academics as a part of geomorphology anyway – ecosystems. The ecosystem concept is not new, dating back over 40 years, but its central position in physical geography has come about only in the past 10 years or so, no doubt as a result of the application of systems thinking in general. If one thinks in terms of ecosystems then one is thinking in terms of systems of varying scales that are influenced by a wide range of inter-related factors. Among other things, it provides a framework for assessing biological factors and the hand of man. That is exactly what the applied geomorphologist has to do, so the ecosystem is a vital concept in applied geomorphology.

The second external development is a general concern about conservation. This is a vast subject. Whole books have been written about it. The applied geomorphologist, indeed the geographer, is only one of many professionals involved in the decision-making process. Nevertheless, the fairly recent and sudden increase in interest in conservation issues has added a new dimension to the applied geomorphologist's role in environmental management.

The third development is the accelerating knowledge of, and exploitation of, the Third World and remote areas of the developed world. Applied geomorphologists are frequently called upon to assist in the investigation of the resource potential of an 'undeveloped' area in the Third World and, to judge from the frequency with which they appear in the media and in academic articles, hazards are affecting Third World countries more than ever before. One wonders whether the hazards are occurring more frequently, or whether they are merely being reported more often, or whether man is inhabiting hazard-prone areas in greater numbers, but whatever the reason it is all grist to the mill as far as the professional applied geomorphologist is concerned. To put it crudely but simply, much of the business lies in the Third World.

The final external development is man–land interaction. The idea that man and the land form a mutually interacting system is an attractively simple one, and at various times it has been put forward as the central, crucial, and distinctive theme in geography. I do not propose to go over the history of intellectual debate on this one topic, but recent thinking is of interest. As recently as perhaps 1980, the consensus among geographers was that geography was gradually but inevitably dividing into two separate subjects – 'human' geography was being absorbed into the social sciences, 'physical' geography into the earth sciences – the implication being that the one had little contact with, and use for, the other. Now there is a swing back the other way. The unity of geography is being stressed again, with man–environment interaction as the core idea. This is an important theme, for example, in the well known Geography 16–19 project, and it is already having an impact on examination syllabuses. If this approach develops strongly, as seems quite likely, it should assure the future of applied geomorphology (man–land interaction) at least for a while. Actually, interesting though this is, it is unlikely that the future of applied geomorphology relies solely on trends such as this. Its practical value and its interdisciplinary nature surely places it beyond the reach of intellectual fashion.

Conclusion

This chapter has made some introductory comments on applied geomorphology. The history of applied studies has been reviewed, the need for applied studies explained, and some contemporary developments described. The next chapter turns to the first major theme in applied geomorphology: hazards.

12 Environmental hazards

Introduction

One of the situations that is of concern to the applied geomorphologist is that in which geomorphological events have a direct impact on man. To some extent man everywhere is influenced by geomorphology. Obviously, the geomorphologist is not involved in every conceivable situation, but he is, or can be, involved when the geomorphic event is of sufficient intensity to constitute a hazard to man.

The general topic of environmental hazards is discussed in various textbooks such as White (1974), Bolt et al. (1975), Waltham (1978) and Perry (1981). In this book, the discussion will be limited to geomorphological hazards, but of course not all hazards are geomorphological. Many others, such as hurricanes and blizzards, are meteorological, and even though they may lead directly to a geomorphological event, such as sea floods or avalanches, they will not be considered as such here. It must also be appreciated that intense geomorphological events constitute a hazard to man only if he is living there or if he is trying to manage the environment there in some way. Many of the earthquakes that occur beneath the sea, for example, pass virtually unnoticed by man. A final introductory point is that while some hazards, such as virtually all earthquakes, are entirely natural, others occur partly because of the activities of man himself. It often happens, for example, that man is partly to blame in cases of flooding or soil erosion. In such cases one usually finds that his interference has led to natural processes acting with increased intensity.

In this chapter the main geomorphological hazards will be described and the effects that they have will be indicated. The great range of geomorphological hazards is often not appreciated, so the hazards are listed for reference in Table 12.1. The catastrophic effects they can have are often not appreciated either, so the world's great natural disasters resulting from geomorphic events are listed in Table 12.2. Loss of life is not necessarily the most meaningful indicator of the magnitude of a disaster, but the table does show that some of our own British examples that appear frequently in the literature seem rather trivial on a world scale.

Soil erosion by water

The natural meaning of the term 'soil erosion' is the erosion of the soil by various natural geomorphological processes. This, however, is usually called 'geological erosion', and is different from soil erosion, which is the erosion of soil that arises when man alters the natural system by various land-use practices. Except in certain circumstances, man himself is not physically removing the soil, but his mismanagement of the land allows natural processes to act with added vigour. Soil erosion, therefore, is accelerated erosion.

Soil erosion by water involves two separate events that occur one after the other. The first is detachment of particles, which is the result of raindrop erosion. The second is the transportation of those particles, and this is achieved by runoff erosion, which might be unconcentrated sheetwash or concentrated flow leading to rill, gully and pipe erosion (Fig. 12.1). The distinction is of practical as well as academic importance. For example, terracing, a common conservation practice, controls transportation but not detachment (Fig. 12.2).

There have been several attempts to establish which parts of the world experience the highest rates of soil erosion. Although the results are inconclusive, it does seem that high rates occur in semi-arid areas and areas with highly seasonal rainfall. However, the impact of man is a crucial factor affecting soil loss. Soil erosion is then important in a number of situations:

(a) Mismanagement of both arable and grazing lands.
(b) Forested areas after firing or tree-felling, and the construction of access roads.
(c) Vegetated land if laid bare, for example, by burning grassland.
(d) Preparation of sites in urban development.
(e) Mining.
(f) The construction of roads.
(g) Where an area is over-used as a recreational facility.

Perhaps understandably, most studies of soil erosion have concerned agricultural land. In this context it is now established that soil erosion is largely determined by factors which relate to climate, topography, soil characteristics, vegetation cover and land use practices. However, vital though this is, much modern work deals with the less familiar topics in the list above. For example, footpath erosion in over-used parts of upland Britain is now attracting much attention, and this is interesting because in discussing soil erosion most people don't immediately think of recreation, nor do they think of Britain. However, the story is told in the pages of both academic journals and

Table 12.1 The geomorphological hazards.

Major structural	Destruction of natural materials by weathering	Fluvial	Slope instability
(a) volcanoes	salt weathering	river floods	mass movements
gases	frost weathering	soil erosion by water	soil creep
lavas	insolation weathering	sheetwash	landslides
tephra (solids)	wetting and drying	sheetflood erosion	mudflows
volcanic ash	root wedging	rainwash	debris flows
lapilli	colloidal plucking	rill erosion	debris avalanches
volcanic blocks	hydration	gully erosion	earthflows
volcanic bombs	hydrolysis	erosion of river beds and banks	loess runs
nuées ardentes	oxidation	changing position of river courses	debris slides
landslides	reduction	sedimentation of river	rock slides
mudflows and lahars	solution	channels and reservoirs –	rotational slips
floods	carbonation	can cause flooding	slumps
	chelation		rockfalls
(b) earthquakes	biological chemical changes		scree movements
ground shaken	includes weathering accentuated		bog bursts
fissures and faults	by pollution		subsidence
slope instability			reactivation of fossil
groundwater circulation			instabilities
disrupted			side-effects of landslides
subsidence			e.g. flooding
seiches			
avalanches			
tsunami			
fire			

Coastal	Arid and semi-arid	Periglacial	Glacial
sea floods	deflation	ground ice	crevasses
tsunami	soil erosion by the wind	permafrost	subglacial and englacial
cliff erosion	sand seas	frost-shattering	caves and meltwater streams
slope instability	reactivation of fossil sand seas	frost-heaving	glacier surges
beach destruction	sand encroachment	frost-cracking	shifting positions of glacier snouts
deposition – in harbours,	desertification	solifluction	and meltwater streams
spit development	salinity	thawing	deposition – meltwater debris
	– affecting agricultural soils	settling	floods, caused by very high
	– affecting building	disruption of frozen ground	meltwater discharges
	foundations	river floods (spring meltwater)	avalanches and avalanche winds
	flash floods		
	debris fans		
	mudflows		

Table 12.2 The mega-disasters.

Volcanoes

1902 Mont Pelée, Martinique, West Indies. St Pierre destroyed: all 30 000 inhabitants killed except for just two survivors.

1919 Lahar. Kelut volcano, Indonesia: 5300 killed.

1985 Nevado del Ruiz, Colombia. Ensuing mudflows killed 23 000 in town of Armero.

Earthquakes

1556 Shensi province, China. Triggered loess landslides and floods: 830 000 peasants killed.

1755 Lisbon: 60 000 killed.

1923 Sagami Bay, Japan: 250 000 killed.

1976 Tangshan, China: 650 000 killed.

1978 Tabas, Iran: 11 000 killed in one town out of a total population of 13 000, and a further 15 000 killed elsewhere.

1980 El Asnam, Algeria: 20 000 killed.

Tsunami

1876 Bengal, India: 200 000 killed.

1883 Java and Sumatra (after Krakatoa): 36 000 killed.

1896 Japan: 27 000 killed.

Mass movements

1618 Landslide, Mont Conto, Switzerland: 2430 killed in village of Pleurs.

1920 Kansu, China. Loess landslips, triggered by an earthquake, blocked river courses and caused floods: 180 000 killed.

1963 Landslide into Vaiont reservoir, Italy, causing floods: 1900 killed.

1970 Debris avalanche, triggered by earthquake, Yungay, Huascaran, Peru: 25 000 killed.

River floods

1887 Hwang-ho, China: 2 000 000 killed.

1931 Hwang-ho, China: 3 700 000 killed.

1970 Ganges delta: 1 000 000 killed.

1978 North-east India: 2800 killed.

non-academic magazines. To quote just one example, Morgan and Scoging (1981) describe soil erosion of this type in the Moel Famau Country Park in the Clwydian Hills (Fig. 12.3).

The effects of soil erosion are not confined just to the area of soil loss. There are considerable repercussions through the landscape. River channels aggrade and become increasingly susceptible to flooding, reservoirs are silted and water quality deteriorates.

Figure 12.1 Gully development: valley-side gullies cut across varied rock types in Coverdale, Yorkshire (from Gregory & Walling 1973).

Figure 12.2 Padi terraces in Luzon, Philippines (Camera Press photograph).

Figure 12.3 The Iron Age hillforts and Offa's Dyke footpath attract many walkers to Moel Famau Country Park. The destruction of vegetation by trampling widens footpaths, resulting in erosion. Geomorphological surveys aid management plans.

Soil erosion by wind

Perhaps surprisingly in view of the conservation movement that followed the North American 'dust bowl' crisis, this problem is still widespread. It is especially prevalent in arid and semi-arid areas where wind velocities and evaporation rates are high, precipitation is low, and drought is common, and where inappropriate development has accompanied recent settlement or where there is some measure of overpopulation. This includes parts of the Great Plains, the steppes of western Siberia and Kazakhstan, the fringes of arid Africa, and parts of India and Australia. It also occurs in temperate areas which suffer periodic drought, and eastern England is now a well documented example (Cooke & Doornkamp 1974, pp. 70–2).

There are several aspects to the hazard. During dust storms, roads may be blocked, ditches and canals filled, fences buried, the dust inhaled, and weeds and insects widely dispersed. Damage to crops can be extensive. Seeds may be removed, plant roots exposed and leaves blasted. The soil is depleted of the more easily removed particles, and this includes smaller grains and some organic matter. This progressively reduces the soil's productivity.

Wind erosion has been the subject of much research, and there are basically three approaches to its study. These are laboratory simulations of the process, such as wind-tunnel experiments, the monitoring of soil losses in the field, and the prediction of wind erosion. All are discussed in detail by Cooke and Doornkamp (1974).

In some parts of the arid and semi-arid world, sand is a genuine threat. There are three types of area, and three types of problem. The first occurs in the great sand seas. When used for nothing more than extensive pastoral nomadism and simple sedentary agriculture at oases, they pose no threat. But blowing sand and even active dune movement can be a hazard when man attempts to 'open up' the area by, for example, building irrigation canals, pipelines, roads, airports, mines, houses and towns. An example of this, and of the practical role that can be played by the applied geomorphologist, is that of a new airport being planned at Dubai (Fig. 12.4) when a team of geomorphologists led by Doornkamp from Nottingham University was called in to assess the dune hazard and to advise on necessary stabilisation measures (Warren & Mainguet 1982). The second type of problem is moving sand on the so-called 'fixed' sand seas. These lie outside the present deserts. They are fossil dunes that were deposited when the deserts expanded beyond their present limits in the late-Pleistocene and were later fixed by grasses and trees. In these areas, plants have only a tenuous hold that can be lost if there is intense grazing, trampling or cultivation. The third problem is desertification. This is

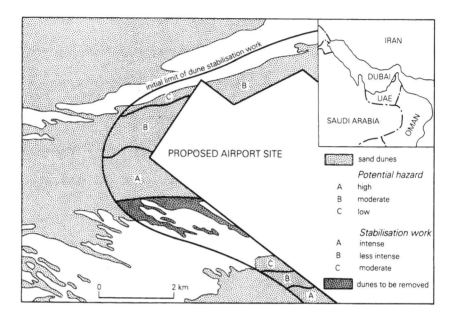

Figure 12.4 The dune hazard at the proposed Dubai Airport.

the progressive deterioration of climate, ground conditions and vegetation towards increasing aridity. It affects mainly the semi-arid areas that lie on the borders of existing deserts. As desertification sets in, the semi-arid areas turn into deserts. Their thorn scrub vegetation thins to the point where the wind can remove the fine soil to leave a surface of bare rock or pebbles. Elsewhere the sand is deposited: a continually moving surface has been created.

Desertification can be caused by natural climatic change towards aridity and by ill judged activity by man. In the African Sahel it is the result of a combination of both. To the effects of the drought in the Sahel between 1968 and 1973 (Fig. 12.5) must be added the effects of man. Typical of a Third World area, there is a very high rate of population increase of about 3 per cent per annum. This has several implications.

(a) *Over-use of woody plants for cooking fuel.* The 100 million people who depend on wood to cook in Africa and the Middle East alone must destroy about 25 million ha each year, even allowing for the ability of the environment to regenerate woody tissue.

(b) *Over-grazing.* Most of the people are pastoralists, and livestock numbers increase roughly in proportion to numbers of people. Stock densities come to exceed the carrying capacity of the land.

146

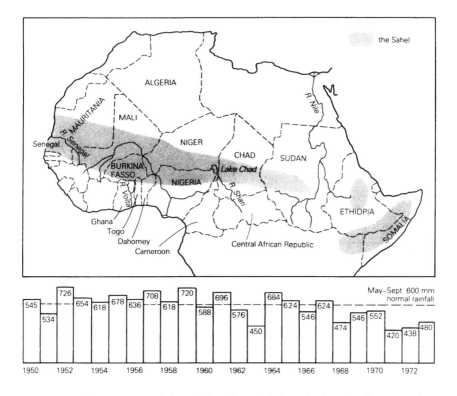

Figure 12.5 The countries of the Sahel, and a rainfall graph showing divergence from average conditions (from Reed 1979).

The plant cover is reduced, the ground trampled and compacted.

(c) *Over-cultivation.* In response to overgrazing, and also partly to political changes depriving them of their traditional nomadic routes, many pastoralists have turned to sedentary cultivation.

(d) *Water supply.* To relieve the water supply problem, boreholes can be drilled to tap groundwater supplies. This has created a problem. It has led to great increases in numbers of livestock irrespective of the capacity of the land to support them. The vegetation is destroyed over wide areas around each borehole. In the 1968–73 Sahelian drought, most animals died of hunger, not thirst.

(e) *Loss of rainfall.* The various circumstances in which vegetation is removed lead to an increase in the albedo of the soil. Less solar radiation is absorbed by the ground. Therefore, there is less ground heating and thus less convectional rainfall. Removal of vegetation could decrease rainfall in this way by up to 40 per cent.

147

River floods

A flood occurs when the capacity of a river channel is exceeded. It is one of the most ubiquitous hazards to affect man, mainly because he chooses to live in so many flood-prone areas. Indeed flooding is an extremely widespread natural phenomenon, and is a problem in those places where man has elected to use areas susceptible to flooding and where he has induced flooding that would otherwise not have occurred. The problem is one of inundation of floodplains and, in arid and semi-arid areas, alluvial fans. The fans occur usually where ephemeral streams from mountains spread out onto adjacent plains. An interesting case study of flooding on alluvial fans in southern Israel and Jordan is given in Cooke and Doornkamp (1974, pp. 125–7). River floods constitute a massive subject, and whole books have been written on this one topic, such as Newson (1975) and Ward (1978).

The dimensions of the hazard are well known. The flood brings increased discharge, high average velocities, high sediment discharges, large individual boulders, inundation of the flood plain, erosion of river channels and damage to farmland or property or both. The loss, in terms of both money and life, can be considerable (Table 12.2). Individual examples seem inappropriate when a phenomenon is so common, but in Britain the flood at Lynmouth in August 1952 strikingly illustrates the effects of flooding. During the storm 20 000 million litres of water fell over the 100 km^2 Exmoor catchment, and the discharge reached 511 cumecs, a figure only twice exceeded by the Thames this century in a basin 100 times larger. About 100 000 tons of boulders were moved into the lower valley.

A number of physical characteristics of floods are important in considering the impact of flooding on man: the frequency of flooding, peak discharge, the total flood runoff discharge, the rate of discharge increase and decrease, the lag time, the area inundated, the velocity of flow, the duration of inundation, the depth of water, the sediment load of the flood, and the time of year when the flood occurs.

A number of factors control these flood characteristics. Taking a world view, the two most common causes of flooding are high-intensity rainfall and snowmelt, so that flooding is a seasonal occurrence in some climatic zones. Ground conditions are also important since infiltration capacity and antecedent soil moisture conditions will affect the conversion of precipitation to discharge. Certain specific events also can cause flooding, for example, the failure of dams, volcanic eruptions beneath glaciers, and the drainage of subglacial and englacial lakes. Several drainage basin characteristics are also significant. For example, other things being equal, high drainage density, absence of lakes and marshes, and circular-shaped basins all

148

promote flooding. Land use changes can also cause flooding. In general, there seems little doubt that many of the floods that occur on mid-latitude rivers would not occur if the vegetation were in its natural state. Some specific case studies have been researched in detail. For example, Howe, Slaymaker and Harding (1966, 1967) show that the increased flood hazard since the start of the century on the Severn in mid-Wales is related to afforestation and to improvements in land drainage in the catchment. Perhaps an even more significant land-use change is urbanisation. Hitherto vegetated rural areas become covered with buildings, concrete and tarmac. Evapotranspiration is reduced, and the generally impermeable surface and network of artificial gutters and drains ensure that a given rainfall is evacuated as efficiently as possible from surface to stream. The result is that flood peaks are increased, lag times reduced, and both the rising and falling limbs of the flood hydrograph are steepened (Fig. 12.6). As a rider, it must be added that man's interference with fluvial systems, through urbanisation, by no means ends there, since we must add to the list increased sediment yields during construction work, increased pollution, abstraction of water from rivers for domestic and industrial purposes, discharge of effluents into them, and a whole range of artificial constructions along the channels themselves.

Slope instability

Slopes have always figured strongly in geomorphological studies, early work searching for all-embracing theories of slope development (e.g. Davis 1909, Penck 1924 and King 1950) and later writers stressing present-day slope form, properties and processes (e.g. Brunsden 1971, Carson & Kirkby 1972, and Young 1972). Contemporary work on materials (e.g. Yatsu 1966, Carson 1971, and Whalley 1976) laid the foundations for slope geomorphology to be able to contribute considerably towards the immensely practical problem of slope instability which is discussed by Zaruba and Mencl (1969) and Voight (1977).

The mass movements that result from slope instability form a continuum of phenomena from soil creep to large landslides. Landslides constitute the main hazard to man and these can be classified into translational slides, rotational slips, and falls. Some landslides are flows; the best-known are mudflows and earthflows, but closely related to them are lahars (volcanic mudflows), bog bursts and loess runs, and all can be a hazard to man, as described by Brunsden (1977).

Gigantic slides have been reported from nearly every mountain area of the world, and almost every year a major disaster is caused by

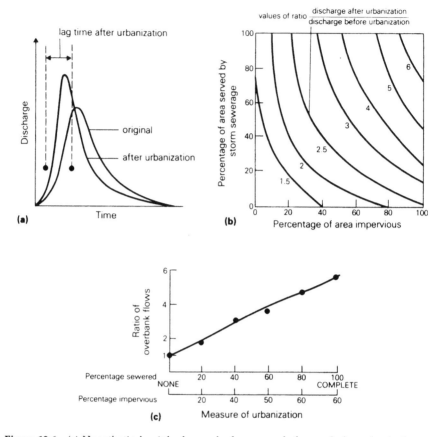

Figure 12.6 (a) Hypothetical unit hydrographs for an area before and after urbanisation; (b) the effect of urbanisation on mean annual flood for a 1 sq. mile drainage area; (c) the increase in the number of flows per year equal to or exceeding channel capacity (for a 1 sq. mile drainage area), as a ratio to the number of overbank flows before urbanisation (after Leopold: from Cooke & Doornkamp 1974).

landsliding. When reported by the media these create great public interest, as happened with the Aberfan coal-tip disaster of 22 October 1966, when 174 people were killed including 146 children in a single school.

There is seldom only one cause of slope instability. Landslides occur because the forces creating movement exceed those resisting it. In effect, it is a movement from an unstable towards a stable state where an equilibrium of forces will be achieved in the form of a repose slope for the landslide debris. The factors encouraging slope instability are many, but they are well understood and are listed in Cooke and Doornkamp (1974, pp.152–3). One point that is clear is that man himself acts as an agent of slope instability, because his power to modify a hillside has been transformed by technological developments. Excavations are going deeper, man-made structures are larger and areas are being used for civil engineering sites which are only marginally suitable. This has become a matter for legislation in Brazil as elsewhere.

Ground-surface subsidence

Although there are several natural causes of subsidence, in terms of environmental management the most important are those attributable to man since the subsidence is often rapid and recent. There are three types of man-induced subsidence. The first, and most important, is the withdrawal of fluids, especially oil and water, which usually leads to slow but extensive subsidence. The second is the rapid and localised subsidence that results from compaction of sediments after irrigation or land drainage. The third, also localised, results from the underground extraction of solids, the most obvious examples of which are the old coalfields of Europe.

Man's traditional response to subsidence has tended to be repair of the damage followed by adjustment to the changed circumstances. Recent technological advances, however, enable environmental managers to retard, prevent and even reverse subsidence, and in modern resource exploitation stress is laid on control of the phenomenon.

Coasts

The coastal environment is characterised by variable and rapid geomorphological processes, and there are four ways in which it gives rise to hazards.

(a) *Pollution.* The discharge of sewage effluent into the sea presents an obvious hazard to the coastal area involved, but what is not so obvious is that, because longshore currents play an important part in the dispersal of such effluent, beaches down-drift of the effluent points are also vulnerable.

(b) *Erosion.* The erosion hazard has three different aspects. The first is long-term retreat of the coast. In some places the rate is extraordinarily high. For example, on the Holderness coast of eastern England long-term rates of retreat average about 1 m per year, enough to have consumed whole villages during historical

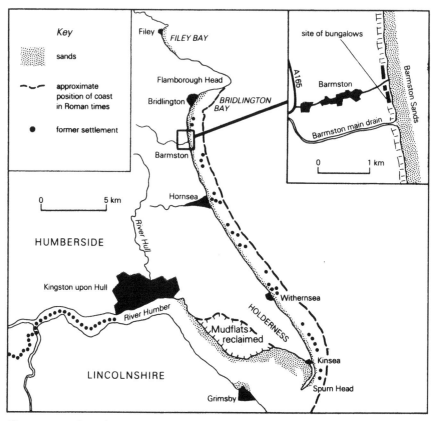

Figure 12.7 Coastal retreat at Holderness, eastern England (from Brunsden & Doornkamp 1977).

times and to leave the area today with a chronic problem (Fig. 12.7). The second is short-term cliff erosion, usually occurring under conditions of destructive high-energy storm waves. De Boer (1977) gives a graphic description of the demise of the village of Barmston on the Yorkshire coast, and the situation of people living in potentially hazardous positions is repeated in many areas. Cliff falls take the form of a variety of slides, slumps and flows, and thus become one specific aspect of the hazard of slope instability discussed earlier. The threat is to residents, property and recreational users. The third aspect is the degradation of beaches. Sandy beaches are extremely vulnerable to erosion under storm conditions. Under some circumstances the foreshore foundation may be lowered so that the beach never regains its former position. Quite apart from the loss of a recreational

152

resource, the erosion of a beach leaves the coast behind more open to attack.

(c) *Coastal deposition.* Deposition leads to less obvious but nonetheless significant hazards. The silting of ports, harbours and navigational channels has implications for trade and industry. A different depositional problem arises when there is accretion of mud in areas used for recreation. This is the case, as described by King (1974, p.218), on the Lincolnshire coast south of Skegness.

(d) *Flooding.* The sea flood is the result of different causes in different parts of the world, for example, tsunami, hurricanes and storm surges. Storm surges are the main cause of disastrous sea floods in Britain. An example is the 1953 flood that affected eastern England. A depression passed to the north of Scotland and then moved south into the North Sea. The position of the low-pressure centre over the North Sea added a northerly gale down the east coast, which, together with the narrowing and shallowing of the North Sea in the south, increased the amplitude of the resultant surge. The surge itself is a pressure effect: sea level rises 30 cm for a pressure drop of 30 mb. In 1953 the surge coincided with high water in the fortnightly tidal cycle, so a 2.75 m surge was added to high tides. There were two disastrous effects. The first was that waves passed over beaches that normally protected cliffs, so that cliffs in, for example, Lincolnshire were eroded. The second was flooding. Extensive areas of the coast were inundated, 300 lives were lost and 300 000 people were evacuated from their homes. London was part of the area affected. London is continually under threat of flooding, and as time goes by the possibility that the flood will be a disastrous one increases. This is because London is gradually sinking, partly because of abstraction of water underground and partly because of crustal downwarping in the southern part of the North Sea area. The Thames barrier, opened in 1984 (Fig. 12.8) is designed to counter this threat (see Hilton 1979, pp. 181–3, Shaw 1983).

Periglacial environments

The hazards presented by periglacial environments can be listed as:

(a) Permafrost.
(b) Frost heave, a vertical displacement measurable in centimetres.
(c) Frost cracking.
(d) Thawing, settling and solifluction, associated mainly with the active layer, and leading to various types of slope instability and

Figure 12.8 The Thames barrier at Woolwich (Handford Photography).

mass movement such as slumping, mudflows, landslides and debris avalanches.

As in other marginal environments, such as tropical rainforests and deserts, these hazards have operated with added impact just recently as technically advanced man has exploited periglacial environments. In doing so, he has modified the natural conditions in three main ways:

(a)   Removal of a vegetation which is very slow to regenerate.

(b) Modification of drainage conditions.
(c) Construction of buildings, roads, pipelines, etc.

The general effect of these modifications is degradation of the permafrost, often irreversible, leading to a disruption of the natural thermal equilibrium in the ground. The changed properties of the ground affect a whole host of man-made structures, such as water supply, waste disposal and drainage systems, roads, railways, runways, telephone lines, buildings, mines, dams, reservoirs and bridges.

Avalanches

An avalanche is the fall by gravity of a mass of ice and snow down a mountainside. It is the main natural hazard in certain mountain areas. Avalanches can occur wherever there are high mountains, open slopes and high snowfalls. They can maim and kill humans and animals, destroy buildings and forests, block highways and railways, and sometimes cause floods by damming rivers. Avalanches carry rock with them and thus can act as agents of erosion. Of the several types of avalanche, the most destructive is the airborne-powder avalanche that can travel at 200 km/hr or more with accompanying destructive blasts of snow-free air. Indeed, the rush of air in front of an avalanche, the avalanche wind, is a recognised additional hazard. Fortunately only a small percentage of avalanches cause death and destruction, but, nevertheless, several do so each year in areas such as the Alps. The worst year in the Alps appears to have been 1951, when a sudden spring thaw followed heavy snowfall.

Such is the desire to protect against the avalanche hazard that an Avalanche Research Centre has been established at Davos in Switzerland. The measures taken in that country have been described by Diem and Conway (1982). The authors divide the course taken by an avalanche into three zones – the 'release zone' on the upper slopes, the 'track' on the mountainside, and the 'runout zone' where the avalanche comes to rest in a valley – and discuss the protective measures appropriate to each zone. In the release zone, for example, the most effective single measure is a stabilising cover of coniferous forest, whereas the runout zone features a wide range of devices and techniques such as vaults, bunkers, tunnels, galleries, sheds and the use of explosives to release the snow prematurely in small quantities. New roads on the Gotthard, Simplon and Grand St Bernard passes have been enclosed in tunnels and galleries for complete protection, while the section of the Trans-Canada Highway through the Rogers

Pass in British Columbia is protected by many miles of avalanche sheds.

Volcanoes

Although strictly a geological topic (Macdonald 1972, Francis 1976, Oakeshott 1976, Rittmann & Rittmann 1976), volcanoes traditionally have been an object of study by geomorphologists. On the land masses, volcanoes are associated mainly with mountain ranges, so the volcanic belts are generally thinly populated; it is in some island-arc areas such as Japan and Java where higher population densities are involved. In any case, volcanoes present a hazard to more than just the agricultural communities that cultivate their fertile slopes, as shown by the destruction of Pompeii and Herculaneum by Vesuvius in AD 79 when 2000 people were killed; and by the threat that Kafla poses today in Iceland to a nearby factory and power station (Escritt 1980). Volcanic eruptions and their consequences vary enormously, but in general the hazards can be listed as follows.

(a) The initial explosion.
(b) Lavas and some other liquids. These rarely pose the main threat because they flow slowly.
(c) The fragments blasted out of a volcano, correctly called tephra. Material of sand size or finer is called volcanic ash, other fragments up to 6 cm are called lapilli, and the largest material consists of volcanic blocks (broken fragments of solid rock) and volcanic bombs (clots of liquid lava that cool and solidify in mid-air).
(d) Gases, mainly water vapour, but including many others. Perhaps surprisingly, searing hot gases caused almost all the 30 fatalities after the eruption of Mt St Helens in 1980 (Francis 1980).
(e) Nuées ardentes, or glowing avalanches. These are masses of hot gases, superheated steam and incandescent volcanic dust that flow down the slopes of the volcano at high velocity. Fortunately glowing avalanches are not associated with every volcano, but when they do occur they are devastating. Forests and other objects in their path are burned and flattened. The infamous eruption of Mt Pelée on the Caribbean island of Martinique in 1902 resulted in a glowing avalanche that engulfed the town of St Pierre. All but two of the population of 30 000 were killed by breathing in gases at nearly 1000°C, making this the worst volcanic disaster ever in terms of loss of life (Fig. 12.9).
(f) A number of side-effects. These include heavy rain; various types

Figure 12.9   The ruins of St Pierre after the 1902 disaster, with Mont Pelée and its spine in the background (after Lacroix: from Waltham 1978).

of debris flow such as landslides, hot and cold avalanches and hot and cold mudflows; the melting of summit snowfields that lead to flooding and sedimentation in reservoirs; and flooding caused by volcanoes erupting beneath glaciers. This last side-effect is a hazard particularly in Iceland. Grimsvötn lies beneath Vatnajökull and Kafla beneath Myrdalsjökull. An eruption of Kafla in 1918 maintained a meltwater flood for two days with twice the average discharge of the Amazon.

Volcanoes cannot be controlled, and man's reponse focuses on understanding volcanoes so that they can be predicted, both in the long term and the short term, and so that endangered areas can be evacuated or avoided altogether.

Earthquakes

Thousands of earthquakes occur each year. Fortunately most take place harmlessly beneath the sea, and very few occur with sufficient intensity on land in densely populated areas to cause disasters. Nevertheless, damage to property and life can be considerable. Reliable statistics are hard to come by, but Davies (1982) estimates that over 1.25 million people have died in earthquakes since 1900. Certain earthquakes rank as some of the worst natural disasters of all time in terms of loss of life (Table 12.2). In 1556 one earthquake claimed the lives of about 850 000 peasants in the Shensi province of China, partly the result of the ensuing landslides and floods, and in 1976 a total of 655 000 died in an earthquake centred in Tangshan in China. In 1980 there were two disastrous earthquakes only one month apart: at El Asnam in Algeria (20 000 killed) and in southern Italy.

The effects and side-effects of earthquakes depend on their location and intensity, but in general can be classified as follows.

(a) *Rock movements,* creating faults of varying horizontal and vertical displacements.

(b) *Shock waves.* The ground is thrown into ever-changing undulations up to 0.3 m high. Fissures open. The shaking of the ground also causes the liquefaction of soft sediments such as alluvium on which buildings are often sited, and the buildings can sink and subside as if into quicksand.

(c) *Slope instability.* Various types of slope instability such as landslides, landslips, slumps, mudflows and avalanches. In turn, these can dam rivers temporarily and cause flooding.

(d) *Subsidence.* Ground-surface subsidence.

(e) *Groundwater.* Groundwater circulation is disrupted.

(f) *Seiches* – waves formed by ground movement in lakes and rivers – which can result in water being slopped over lake rims or dams.

(g) *Tsunami.* These are giant sea-waves induced by earthquakes. They can be as much as 60 m high and are much more destructive than other sea floods. They rate as major natural hazards in their own right. In 1883, 36 000 inhabitants of the low coasts around Java and Sumatra were drowned in the wake of Krakatoa, and tsunami in Japan in 1896 claimed 27 000 lives.

The scene after a violent earthquake in a built-up area is one of devastation. Bridges collapse, roads are fissured, and railways are buckled and twisted. Water and gas pipes are cracked open, and other services are dislocated. Fires often break out (due mainly to broken gas mains and electrical short-circuits), disease is a threat, and relief work is hampered. Most deaths, however, result from the collapse of houses. According to Davies (1982), 80 per cent of all earthquake deaths occur in this way, and the most vulnerable buildings are those which have flat roofs covered with a thick layer of mud.

Man has limited defences against the earthquake hazard. Further, the devastations of, for example, Lisbon in 1755, Tokyo and Yokohama in 1923, and San Francisco in 1906 act as reminders that some highly populated areas are under continuous threat. Although man-made structures can be built specially to withstand earthquakes, man has little or no control over earth tremors and there is little time for precautionary measures. It is a hazard he must learn to live with.

Earthquake is the subject of textbooks by Eiby (1967), Oakeshott (1976), Wiegel (1976) and Bolt (1978).

Geomorphological hazards case study: the Karakoram

Certain parts of the world are unfortunate in that they experience an unusually large number of natural geomorphological hazards. Such a region is the Karakoram, a mountainous area in northern Pakistan (Fig. 12.10). In the summer of 1980 the region was the object of study for a team of 70 scientists from China, Pakistan, Switzerland and Britain, an expedition organised by the Royal Geographical Society and called the International Karakoram Project. Some of the findings have been published in three recent *Geographical Magazine* articles by Goudie (1981b), Wood and Hunt (1981) and Davies (1982).

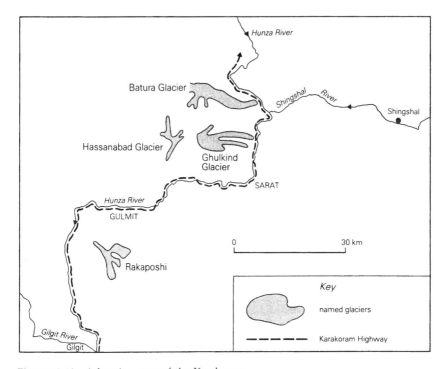

Figure 12.10 A location map of the Karakoram.

159

The Karakoram has the following combination of natural physical features.

(a) A vertical height range between valley floor and mountain top that is unequalled anywhere on Earth. At one point, for example, Rakaposhi (7772 m) towers a massive and unique 5990 m above the Hunza Valley (1800 m) only 8 km away.
(b) The glaciers are the largest in the world outside the polar regions.
(c) The Karakoram is a desert. Rainfall averages 10–12 cm per year, and summer temperatures frequently exceed 40°C. The International Karakoram Project witnessed seven severe duststorms in six weeks.
(d) The area is subject to earthquakes.

In spite of these adverse characteristics, the valley floors are inhabited by peasant farming communities, and in recent years there have been improvements in communications. The Karakoram Highway, which links Pakistan with Tibet, follows the valleys of the Indus and one of its tributaries, the Hunza. It goes through some of the most difficult terrain in the world and was built in the face of massive construction problems. A second example of improvements in communications is the path built in 1966 along the Shingshal Valley, a tributary of the Hunza, to the village of Shingshal. It is against this backcloth of habitation and development that one must view the hazards to which the area is subjected. The hazards result directly from the physical features of the landscape. According to Davies (1982) the inhabitants perceive the hazards in the following order of importance.

(a) *Water shortage.* Failure of water supply.
(b) *Land erosion.* This can lead to the abandonment of farmland and even villages. The Karakoram experiences very high rates of denudation – an average of about 5000 t/km^2/year, which is about 100 times more than over the Thames basin. The factors involved are the great available relief, the friable nature of the rock, the intensity of frost and salt weathering, the frequency of rockfalls and landslides, the paucity of vegetation, and the high summer river discharges when the ice and snows melt. In fact, the hazard does not end there, for the sediment passes downstream to the Indus where it builds up behind dams, reducing the lives of the reservoirs to decades.
(c) *Flooding.* These are mainly meltwater and flash floods, but in fact the most horrific floods occur when mudflows and landslides block a river, forming a lake, only for the temporary natural dam to fail. In 1858 a great slab of rock and glacial debris blocked the

Hunza River at Sarat. The blockage gave way 6 months later, releasing such a massive volume of water that the Indus rose 16.8 m in 7½ hours at a site 250 km downstream. In 1841 the release of a lake 55 km long caused a 24 m rise. If such a wall of water were to occur today it might destroy the Tarbela dam.

(d) *Landslides, rockfalls and mudflows.* The degree of relief, the sparseness of vegetation, and the fractured and contorted nature of the rocks make these slopes particularly unstable and threatening. A 139 km section of the Karakoram Highway between Gilgit and Gulmit was observed by the International Karakoram Project to meet 335 debris falls, rockfalls, mudflows and landslides. Parts of the path to Shimshal have tumbled into the river, and others have been overwhelmed by continually shifting screes, some of them over 2000 m high. Further, as indicated above, slope instability can lead to further hazards related to lake creation and flooding. For example, the International Karakoram Project witnessed a mudslide at Gupis that blocked the Ghizar River. This formed a lake 3.3 km long that destroyed several wheat mills, slowly inundated a village drowning 21 houses, destroyed crops, killed 2 villagers, and affected 600 people in all.

(e) *Earthquakes.* In 1974 an earthquake at Pattan killed 1000 people. Davies's article (1982) is concerned mainly with the building designs used by the local people in view of the earthquake hazard. Earthquakes can also trigger off various forms of slope instability discussed above.

There are three points about this list that are worthy of comment.

(a) The sequence selected by the inhabitants is not necessarily the one that the reader would have expected.

(b) The inhabitants explained to the International Karakoram Project that they had far greater worries than any of these hazards. Their priorities were more to do with lack of education for children, their health, the shortage of medicine in the dispensary, and the difficulty in getting a good price for their crops.

(c) There are other hazards in the Karakoram that have not been mentioned so far. They are related to the glaciers in the area. The glaciers create the following hazards:

(i) They can stand in the way of lines of communications. The path to Shingshal, for example, involves a glacier crossing.

(ii) The outwash streams from the glaciers can also obstruct routeways. Between Gilgit and Gulmit (139 km), for example, the Karakoram Highway crosses 18 glacial outwash streams.

(iii) The size of the glaciers can fluctuate considerably. At the turn of the century, for example, the Hassanabad Glacier advanced 9.5 km in 2½ months, a rate unmatched anywhere on Earth, and in 1953 in the same region the Kutiah Glacier moved forward at an estimated rate of 113 m/day. Glacier advances can devastate fields and destroy irrigation channels. Also, in 1974, the Batura Glacier demolished a major bridge on the Highway, because the position of the snout changed abruptly.

(iv) The meltwater streams issuing from the glaciers can also shift their courses unexpectedly. This poses an obvious threat to man, his houses and his farmland. It also affects the Highway. In 1980 the International Karakoram Project observed outwash streams from the Ghulkind Glacier block part of the road that was thought to be safe from this hazard. In fact, the waters carried away a long stretch of road and stopped communications for several days.

Conclusion

The geomorphological hazard is a practical topic, none more so. However, I want to conclude this chapter by making some comments on the way that new developments provide interesting sidelights on old debates.

The circumstances in which catastrophism was replaced by uniformitarianism in the 19th century were described in Chapter 1. Uniformitarianism is the belief that landforms are the product of present-day processes operating slowly over long periods. Catastrophism appeals to the Creation and the Flood. Uniformitarianism is right, of course, not the Flood, but nevertheless certain modern developments in geomorphology suggest that we are now adopting a more catastrophist viewpoint than one might think.

One of these modern developments has nothing to do with hazards. It is the emphasis on climatic change in landform explanation described in Chapter 6. In many cases we are virtually saying that landform origin dates from the Pleistocene. That is fairly 'catastrophic' in view of the millions of preceding years of geological time. Moreover, at least in the mid-latitudes, we are doubting whether present-day processes have formed certain landforms and whether those processes have been operating for more than a few thousand years. It all strikes at the very heart of uniformitarianism.

Hazards have a bearing on this discussion in two ways. First, some of these hazards are the results of crustal structural movements that operate over very short timespans indeed: volcanoes, earthquakes, faulting. Secondly, and perhaps more interestingly, some of those hazards that result from the operation of geomorphological processes also take place over short time-periods – days, sometimes even hours. Examples include river and sea floods and landsliding. The event is short-lived, abnormal, and could be described as catastrophic in more ways than one.

The concept linking these phenomena is the threshold, described in Chapter 10. Hazards occur when a threshold is exceeded – a river floods, a cliff-fall occurs, a slope movement takes place. Viewed in this way, the threshold is also the threshold between uniformitarianism and catastrophism. Most of the time the landscape is being acted on by long-continued processes that are of moderate intensity. This is the idea encapsulated by uniformitarianism. Then suddenly the rare, high-intensity geomorphological event causes a threshold to be breached, and we have a hazard, and we are speaking in catastrophic terms. Therefore, by putting hazards in applied geomorphology to the forefront of the subject we are re-emphasising, albeit in a different guise, the discredited old paradigm of catastrophism.

13 Environmental management

Introduction

Man has always had some geomorphological impact on the landscape. It is clear, however, that the scale and pace of this impact has increased as time has gone by, so that most of his remodelling of the land surface has been achieved in just the past 10 000 years and much of it since the Industrial Revolution. This is not simply because of his rapidly increasing numbers, but results mainly from his social organisation, the development and use of tools and the harnessing of energy resources, not just his own energy, but that of animals, water power, machines and explosives. Man the technological animal has become man the geomorphological agent. There seems little doubt that man is now the most powerful geomorphological agent, certainly in Britain and possibly over the world in general.

As pointed out by Jones (1977), man's geomorphological behaviour is unusual in three ways. First, it is controlled more by economic restraints than by environmental factors. Man has the technology to build virtually any landform in virtually any environment. No other process can claim that. Secondly, his activities extend beyond the land areas to the oceans. Thirdly, he can choose what his impact will be – its timing, location and magnitude.

Man's various types of impact on the landscape can be classified in a number of different ways. One can contrast ways in which he has totally altered the landscape – created new landforms – with ways in which he has merely modified it. One can contrast ways in which he has intervened intentionally in geomorphological systems with those where his impact was unintentional. One can contrast ways in which he has directly altered or created landforms with those where he has intervened indirectly by affecting processes. The last two types of classification are popular in the literature, and to some extent they are related. This is because, in general, man's intentional impact has led to the direct alteration of landforms, whereas his unintentional impact has been more subtle via his effect on processes.

Man alters landforms directly by excavating and piling up earth, reclaiming land and causing subsidence. For example, the disturbances created by the extraction of mineral resources (sometimes called land scarification) include open-pit mines, quarries, sand and gravel pits, strip mining and various types of waste tip. The construction of roads and railways also provides several examples: embankments, cuttings and sundry other man-created slopes. There are also examples from agriculture: lynchets, terraces, drainage ditches and irrigation channels. It will be noted that most of these effects are intentional as well as direct.

The indirect effect on landscape via process is more widespread and more important. It represents a natural readjustment of the geomorphic system where the equilibrium has been upset by man. Again there are many examples, but probably the most important situations arise when vegetation is cleared – soil erosion, river floods, increased sediment discharges, and so on. Here, of course, the effects were not intended. Sometimes man deliberately builds a structure which yields unintentional results. For example, the creation of a reservoir can lead to slope instability and, downstream from the dam, channel erosion. An even more obscure example of the indirect effect is that man's unintentional modification of the atmosphere will affect processes which in turn affect landscape development.

One group of situations where man deliberately acts as a geomorphological agent is special. It is where man intervenes with an informed concern for the environment. This is environmental management, and it is the subject of the rest of this chapter. There is one essential prerequisite: man must understand the operation of the relevant geomorphological systems, otherwise he is not managing the environment, just tampering with it.

There are two aspects to environmental management. The first is man's planned and carefully judged responses to the existence of the various geomorphological hazards. The second branch concerns man's alteration or conservation of the landscape in such a way that he gains some sort of benefit; for example, he can take measures to conserve a beach that he wants for his own recreation. Therefore, some of the following sections correspond to some of the hazards discussed in Chapter 12, while others deal with man's manipulation of the environment for his own ends. There are areas of overlap, however, between the two branches of environmental management, because in constructing something that he wants, man has to take account of potential hazards. For example, if man builds a motorway, that is not a response to a hazard, but in this particular piece of environmental management he will have to be aware of some possible hazards such as slope instability.

The remainder of this chapter, therefore, includes both these types of environmental management, and in doing so covers numerous examples of man's intentional, unintentional, direct and indirect effects on landscape.

Control of soil erosion by water

In Chapter 12, seven different situations in which soil erosion occurs were identified (p. 139). The different situations require different conservation measures. This chapter will consider only one of the situations: soil erosion from agricultural land. Much research has been done on it, and it is one of the problems hampering increased world food production.

Conservation practices fall into three groups. The first is the practice of crop management – the use of a legume or grass crop in rotation at least one year in five, judicious applications of fertiliser and manure, cover crops, intercropping, mulching and so forth. The second comprises practices that support the crop management ideas – contour farming, strip cropping and terracing. The third concerns the practices designed to restore eroded land – a protective natural cover of vegetation can be encouraged, gullies can be converted into stable artificial channels of appropriate dimensions, the water supply to gullies can be reduced by conservation practices in the tributary areas, the flow of water in a gully can be stopped entirely by diverting it into an artificial channel, and so on.

Most readers will greet this list in the same way, a feeling that it is old hat, that they have heard it all before. Therein lies the danger. Soil erosion is a familiar, long-standing hazard. It dates back at least to the start of this century. And it is less prominent in environmental debates than it used to be. Perhaps, as pointed out by Cooke and Doornkamp (1974, p. 21), the notion is widespread that because conservation practices have been developed to combat soil loss, the problem has been solved. This is not so, and for three reasons.

First, conservation practices have not always been successful. A 1977 Food and Agriculture Organization report admits that many mistakes have been made, often leading to a worsening of the situation. Soil erosion has been accelerated by, for example, a bad choice of cropping patterns, unsuitable cultivation techniques, implements and machines, the misuse of tractor power and the extension of cultivation to marginal and submarginal land. Perhaps the basic mistake has been the careless use of Western techniques in the fragile environments of the tropical world.

Secondly, in many areas of soil erosion, conservation techniques are not used. There are a number of reasons for this. Some farmers do not see soil erosion as a problem. Also it is difficult to disseminate information to those who require it. Therefore, knowledge about soil erosion must be transmitted to farmers and other resource users in an intelligible and acceptable way: the practices must be adopted. This work is as much a task for the geomorphologist as the physical problem itself.

166

Thirdly, there remain large areas of ignorance among scientists. For example, too little is known about the 'tolerable soil loss' in different environments; basic data on the erodibility of soil is scarce in most countries; and it is important to be able to recognise the symptoms of soil erosion at an early stage before the most valuable surface humic horizons are lost.

The message is clear. There is much to be done, and the applied geomorphologist must maintain his involvement in the management of areas experiencing soil erosion.

Control and prevention of soil erosion by wind

Reduced to its simplest, the requirement is to reduce the effective wind velocity and/or to improve ground-surface conditions so that saltation cannot begin. There are three groups of control measures that help to achieve this.

The first is the various methods concerned with crops and vegetation. This includes erecting windbreaks (about which much detail is known – see Cooke & Doornkamp 1974, pp. 62–4) and a number of field cropping practices such as planting cover crops, strip cropping and the management of stubble and crop residue. The second concerns ploughing practices, and considerable attention is being given to the design of agricultural equipment suitable for soils that are potentially vulnerable to soil erosion. The third is a group of soil-conditioning methods. Soil moisture can be conserved by surface mulching, terracing and irrigation. Erosion-resistant clods can be created by, for example, ploughing soon after rain or by increasing the amount of organic matter in the soil.

As with soil erosion by water, these conservation measures are not always implemented. Farmers often fail to recognise the symptoms of wind erosion or to recognise it as a problem. Sometimes the problem is recognised but there is no desire or motivation to solve it. Not all farmers are aware of the remedial measures and the physical principles underlying them. And finally, even if there is a desire to act, there may be a delay for practical reasons such as financial circumstances. The geomorphologist, therefore, has two tasks: one relates to problems and solutions, the other to implementing the solutions.

Rivers and river channels

The deliberate management of rivers and their channels by man is almost universal in areas inhabited by him. In most of the developed world, and in much of the developing world too, the difficulty lies not in identifying ways in which man has modified fluvial systems, but in identifying rivers and channels that undoubtedly remain in their natural state.

Man uses rivers in a number of different ways: as a source of drinking water and food, as an artery of transport, as a means of waste disposal, etc. The list of ways in which he manages them for his requirements is long but familiar: dams, irrigation, drainage ditches, embankments and reinforced channels, weirs, navigability, extraction of water and discharge of effluents. Sometimes his actions produce side-effects which constitute hazards and therefore need further management by man – using the river for waste disposal leads to pollution, dams produce sedimentation of reservoirs, reducing their storage capacity. The significance of all this is that much is known, from pure geomorphology, about the behaviour of the fluvial system, so that the applied geomorphologist is in a position to be able to advise the many authorities who control the development of rivers on the best management techniques.

Responses to river floods

The action man takes in response to this hazard is inextricably bound up with his perception of it. It seems that, in general, the more advanced societies perceive the hazard as more important than do most Third World societies, and that one's awareness of the hazard decreases with passing time. Beyond this it is very difficult to make generalisations and the subject is very complex. In view of this it is just as well that there is a large body of accurate numerical knowledge about river floods. A number of statistical procedures are used, but the most common is the duration curve, which enables one to say how frequently, on average, a flood of a particular magnitude will occur on a particular river. Engineers can then plan to protect against a specific event. According to Hilton (1979, p. 107), for example, urban flood designs often revolve around the flood that would be expected, on average, every 100 years, while stormwater drainage from streets is geared to the '10 year event'.

Man is in a good position to prevent and control floods, since surface water is more amenable to interference than, say, atmospheric processes. Pioneer work at Chicago by White and his students (White 1945, 1964; White et al. 1958; White 1961) has led to the identification of a number of possible responses:

(a) Bearing the loss.

(b) *Public relief.* Interestingly, one effect of all forms of gifts and aid is to encourage the persistence of inappropriate occupation of flood-hazard zones.

(c) Emergency action, such as evacuation.

(d) *Flood-proofing,* designed to reduce damage to structures and goods within hazard zones.

(e) Land use change and regulation.

(f) *Flood insurance,* the implication of which is that geomorphological and hydrological data are vital to insurance companies.

(g) *Flood abatement and control.* These are probably the most effective ways in which man intervenes to control flooding, and this is where the applied geomorphologist comes into his own. Flood abatement is the alteration of land use in the catchment: an attempt is made to prevent the flood from forming in the first place, or at least to reduce the amount of water reaching the main rivers. Flood control is the building of protective measures along the river designed to control the flood in the channel. The two approaches are not necessarily mutually exclusive, although in the United States they led to the so-called 'upstream–downstream' controversy in which a major issue was the relative merits of small headwater dams and large mainstream dams. Flood control schemes are often just one aspect of a plan whereby a whole drainage basin becomes the subject of an integrated water-resources project involving irrigation, hydroelectric power, navigation, domestic and industrial water supply, recreation, fish and wildlife. Examples are the Tennessee Valley scheme, the Missouri River project, the Orange River scheme in South Africa, and the Indus project in Pakistan. Such integrated basin planning gives practical value to Chorley's (1969) idea that the drainage basin is the fundamental geomorphic unit.

Managing landslides

As far as their effects on environmental management are concerned, landslides are of three types.

(a) Where the slide was not predicted and caused great damage and grief. Aberfan is an example.

(b) Where the threat of a slide is known about and something is being done to control the position. There are many examples of this all over the world, but one specific case described by Cooke and Doornkamp (1974, pp. 150–1) occurs at Upper Boat in South Wales where a group of comparatively new houses stands just below the toe of an active slide, whose movement is being controlled by drainage pipes releasing the high internal pore-water pressures.

(c)   Where the slide occurs along a proposed transport route, or where it potentially influences the site of a planned development.

The role of the geomorphologist in the management of landslides is twofold. The first is a predictive role, analysing landslide-prone areas. This is more than a matter of present-day landforms, geology and processes. It also includes past processes. For example, considerable difficulties were encountered during the construction of the Sevenoaks bypass in south-east England because fossil solifluction forms were reactivated (Skempton and Hutchinson 1969). The second role is to identify the controlling characteristics of known landslides and to liaise with the civil engineers in controlling the causes of landslides. It is important to distinguish between those physical characteristics of a site which make sliding possible, and the actual trigger mechanism which initiates the movement. Of the many trigger mechanisms the most common is an increase in pore-water pressure. This can be relieved by adequate drainage of the landslide, and diversion channels can be placed above the head of the slide to prevent surface runoff onto the slide itself.

Coasts

Man's interference with coastal systems is of two types – intentional and unintentional – and his management of them is also of two types – active and passive. There is a link between the two. In intervening intentionally, man is doing things such as protecting the coast from erosion, conserving a beach, dredging a harbour, or reclaiming land from the sea. He is thus adopting an active style of management. On the other hand, management can be passive. Coasts can be left in their semi-natural state and exploited as scenic, amenity and recreational resources; here man's impact is usually unintentional.

Geomorphologists are increasingly involved in coastal management, and it is possible to recognise two main scales of operation. The first is at the site or local level. Here, their role might consist of advice on coastal rehabilitation, monitoring dynamic changes, identifying potential hazards such as landslips and, increasingly important, compiling environmental impact statements. In the United States such statements are mandatory for all federal and many state developments. To some extent the need for them arises because of the many conflicting uses to which coasts are put. This is particularly so with estuaries, and from a management point of view estuaries have been studied by politicians, economists, planners, and ecologists, and, regrettably, the ecological implications are often the last to be considered. The second scale of involvement is at the regional level, and here the role is mainly of terrain analysis and resource inventory compilation and appraisal. One example of this is that in Scotland a survey of all the beaches and dune areas has enabled planners to formulate management strategies that avoid piecemeal developments.

The work of the geomorphologist, of whatever type and whatever scale, is ultimately integrated into a management policy. Most of the active management is concerned with countering the coastal hazards described in the previous chapter. The most important aspect of this is coastal protection and erosion prevention. There are basically three ways in which this can be done. The first is beach improvement. The best protection for any coast is a high, wide beach. Beaches can be improved either by sand by-passing (King 1974, p. 214) or the dumping of material of suitable grain size on the foreshore. This is most successful where the fill is coarser than the natural sand and where the area is fairly sheltered. In these conditions, fill can be obtained from offshore borrow-pits. King (1972, pp. 25–9) describes several examples of where such beach nourishment has been carried out. The second method of coastal protection is to build solid sea-defence structures such as sea-walls, breakwaters, jetties and groynes. To the layman, and indeed to some engineers, this seems to be the obvious solution, but the position is complicated and there are many examples of where such action ultimately aggravated the problem. One from the United States is discussed by Ritchie (1981), and this particular article is interesting also in that it shows how geomorphologists can gradually make engineers change their minds by bringing the results of pure geomorphological research to their notice. A familiar example from Britain is the implication of building groynes. Groynes check longshore drift and therefore are beneficial to the immediate hinterland in that they raise the level of the foreshore, but they can lead to coastal erosion downdrift because the movement of material there is being impeded deliberately. This is happening, for example, at Rottingdean in Sussex, downdrift of the groynes on Brighton beach. In this sort of situation there is a role for laboratory studies, and as an example one can cite the model experiments carried out by Hydraulics Research at Wallingford to assess the optimum pattern of groynes for reducing serious erosion along the Suffolk coast at Dunwich. The third approach to erosion prevention is the stabilisation of sand dunes and their artificial creation. Although this is not always desirable, it is an important form of protection on many low coasts. In Lincolnshire, for example, marram grass has been planted, brushwood fences erected to trap the sand, and efforts made to prevent unwanted human

171

interference. In Holland, dunes have been reinforced to make dykes.

Geomorphologists are also involved in harbour work, the deepening of estuaries to form ports, and the definition of navigation channels. Here, too, there is always the possibility of unwanted side-effects. King (1974, p. 212) quotes the example of erosion that followed the opening of an inlet for navigation at Thyborøn in Denmark.

The final aspect of active coastal management relates to land reclamation. This is most common on low-energy coasts, and good examples are in the Wash in eastern England and along the coasts of Holland and Denmark.

Coastal management must also cater for the use of coastlines for recreation. This includes sailing, surfing, underwater sport, fishing, marinas and beaches. The conservation of beaches is linked to the earlier discussion on beach improvement, for a beach designed for protection is also a beach for recreation. The optimum sand size for holiday beaches is moderate- to fine-grained; a relatively flat beach is produced, with a wide expanse of foreshore.

Coastlines are also the sites of resources, both organic (e.g. seaweed, shellfish, lobsters, crabs) and inorganic, such as sand and gravel. Sand and gravel for constructional purposes are available on beaches and have been exploited. As discussed earlier, however, it is now realised that beach material is vital to protection and conservation schemes, so contractors are now looking to the offshore zone. Under some circumstances the sorting action of waves concentrates heavy minerals, including gold, forming rich reserves on sandy beaches. These have been located on present-day beaches, and on raised beaches (e.g. in western New Zealand), and on submerged beaches (e.g. off Alaska) where they have been exploited.

Periglacial environments

The rational management of periglacial environments involves the geomorphologist in two main ways. The first is terrain evaluation: a preliminary reconnaissance of extensive areas, often using remote-sensing imagery, followed by detailed site investigations. Landforms are important, reflecting as they do surface and subsurface processes and conditions. The second is practical decisions to be made by engineers. The main problem, and the source of all the others, is permafrost, and there are four possible responses to its existence. One response is active: the permafrost can be eliminated. Three are passive: it can be neglected, or preserved, or structures can be designed to take expected movements into account. These various approaches are discussed in detail by Cooke and Doornkamp (1974, pp. 245–7). Two interesting examples of the involvement of geomorphologists in important decision-making are the planning of a route for the trans-Alaska oil pipeline (Fig. 13.1), and the selection of a site for the Canadian town of Inuvik (see Cooke & Doornkamp 1974, pp. 247–51).

The destruction of natural materials by weathering

Rock is the parent material of soils on which food production depends; it is widely used for buildings, roads and other engineering structures, and forms the natural foundation for most of them. Therefore, geomorphologists, pedologists, building-research workers and engineers share a common interest in the various weathering processes, the performance of construction materials when subjected to them, and the changes in rock and soil properties arising from them.

The problem can be approached from two different directions. One is to study weathering processes, the other to study properties of materials. Of course, the two are continuously interlinked, and results must be integrated to give a complete picture.

There is much important recent work on the weathering system, decomposition by chemical action, and disintegration by crystallisation processes. For example, it transpires that, in Britain at least, the most important pollutant is sulphur dioxide: sulphurous gases contribute to weathering by promoting chemical weathering and, even more important, leading to the formation of salts that become involved in salt weathering. Indeed, salt weathering turns out to be a very significant practical subject. It is a hazard particularly in arid environments. For example, in the Suez city development area, damage to buildings and their foundations results from capillary rise of saline groundwaters. Some building materials are affected more than others. A surface resources survey demonstrated that Miocene marine limestone should not be used in construction work because of its susceptibility to salt weathering (Derbyshire & Sperling 1981).

Work on the properties of materials is, in a sense, the practical arm of the contribution of Whalley (1976) and others to pure geomophology discussed earlier. An important distinction can be made between mechanical resistance and chemical resistance. Rocks often have one property but not the other. For example, most limestones have high mechanical strength but are susceptible to solution, whereas slate has only moderate mechanical strength but its high content of mica and other clay minerals makes it very stable chemically under atmospheric conditions. The strength of natural materials is generally closely related to their physical properties, so this opens up the field of predicting the durability of construction materials in the light of knowledge of their properties.

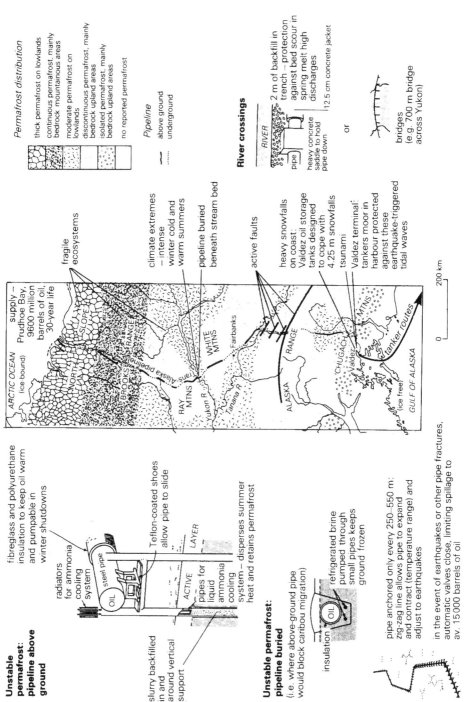

Figure 13.1 The Alyeska pipeline: oil from the North Slope (from Hilton 1979).

One final example will highlight the importance of the geomorphologist. During the site investigation phase in the construction of a reservoir on the Denbigh Moors in North Wales, conventional borehole investigations were inadequate, and the properties of the area beneath the proposed dam could be assessed only by a geomorphologist because of the presence of complicated sedimentary sequences left by an advancing and retreating glacier (Derbyshire & Sperling 1981).

Urban management

Against a background of increasing demand for urban land, much of it in Third World areas where geomorphological knowledge is scanty, a number of aspects of environmental management are brought together here. The geomorphologist has six roles in urban management.

(1) Making initial reconnaissance surveys to select suitable sites for urban development within a region or country.
(2) Mapping potentially hazardous zones within the selected areas.
(3) Giving advice on the beneficial use of topography and surface materials in planning specific routes and layouts.
(4) Assisting with the development of the site, particularly on matters of slope stability and erosion potential.
(5) Advising on weathering problems related to foundations and building materials.
(6) Dealing with post-construction problems, such as analysing hazard damage and monitoring geomorphological changes to reduce later problems.

A good example of the involvement of geomorphologists in urban planning is described by Cooke and Jones (1980). The expansion of the city of Suez, referred to in the previous section, required the prior evaluation of surface resources and geomorphological hazards of which two, flooding and salt weathering, were found to be important.

Road-building

As with other engineering projects, geomorphological surveys are used at three stages in highway construction.

(1) *The reconnaissance stage.* This initial survey should identify:

    (a) The main characteristics of the terrain within which a route corridor is to be placed.

(b) Influences on the route from outside its planned boundaries, for example, floods.

(c) The range of surface materials, since these will be needed for fill, sub-base, base and wearing course.

(d) Foundation problems such as karstification or mining subsidence.

(e) Processes which will affect the construction and safety of the route, for example, slope instability or vigorous gully erosion.

(f) The location, pattern and magnitude of surface and subsurface drainage features, since this facilitates early design and costing of drainage measures.

(2) *Site investigation stage.* Working on the chosen route, the geomorphologist can help to improve vertical alignments and reduce excavation or loading problems, to design preliminary land drainage systems, to provide natural slope parameters to guide the design of cut slopes, to determine the appropriate land-take requirements of the road and for evaluation of budget estimates.

(3) *The construction stage.* Surveys are only rarely carried out by the geomorphologist at this stage since the problems should have been identified long before construction begins. However, he continues in a monitoring and advisory role, especially in the event of problem or failure.

A readable and accessible summary of this subject is given in Brunsden (1981).

Conclusion

Man has always had an impact on the landscape, sometimes intentional, sometimes unintentional, sometimes desirable, sometimes adverse. This impact has accelerated considerably over the past 200 years, and there is no doubt that we must now recognise man as an important geomorphological agent. As man's understanding of the landscape and his understanding of his own effects on it have improved, so he has been able to intervene deliberately with an informed concern for the environment. This is called environmental management.

14 Resource evaluation

Introduction

The previous two chapters have concentrated on the practical relevance of dynamic geomorphic process systems. In this chapter the scene is switched to something more passive – the evaluation of landscapes as resources. Landscapes are at the same time mineral, aesthetic and strategic resources. It is also relevant to consider land systems and geomorphological mapping at this point. This is because landform evaluation as an applied science had its origins in the development of these two mapping techniques during the 1950s, in Australia and Poland respectively.

Material resources

The geomorphologist's contribution towards the evaluation and management of material resources focuses on three topics: soils, minerals, and sand and gravel. Soil is the most ubiquitous resource available to man and its crucial role in agriculture and therefore in world food supply does not need to be stressed. Soils are also central in geomorphology; in most places they are the zone of contact between rock and process, and their important agricultural characteristics are the result of the weathering of parent materials. In the sections on soil erosion, much has already been written about the geomorphologists' role in soil conservation, but equally important is the mapping of soil resources. Here it is common for geomorphological maps produced from aerial photographs to be transformed into soil maps using the concept of the soil catena.

A similar approach can be used in the search for mineral resources. Geomorphological mapping provides the vital starting point. This is because mineral resources are often related directly to relief. Examples include alluvial placer deposits such as gold, diamonds and tin; weathering products such as enriched copper, limonite, manganese, bauxite, cobalt and kaolin; and basin deposits such as coal and iron ore.

In terms of tonnages, sand and gravel are the two most important materials extracted from the earth. They are required mainly in the building and construction industries. Aggregates for concrete are in the greatest demand. There are four areas where the geomorphologist makes a special contribution.

(a) Identifying, mapping and evaluating economically viable deposits, and designating dredging zones.
(b) Preparing environmental impact statements and advising on exploitation techniques.
(c) Monitoring an area during and after its exploitation.
(d) Evaluating the various costs and benefits accruing from mining operations.

The first area is traditionally his most vital role: assessing the nature and distribution of available resources. He is able to do this because suitable sand and gravel deposits nearly always occur in well documented geomorphological situations. These are the channels, floodplains and terraces of rivers; fluvioglacial environments; marine environments (especially on beaches and in the offshore zone); and on some slopes, where suitable materials are to be found, for example, in screes. This intimate theoretical knowledge of the depositional environments of sand and gravel enabies the geomorphologist to play the role of prospector in the aggregate industry. One example of this, discussed in detail by Joliffe and McLellan (1980), is that analysis of prolonged stillstand positions of former ice-fronts, identified viable sand and gravel deposits in central Scotland. Even this one task of prospecting and compiling inventories involves the geomorphologist in a number of jobs: preliminary survey of possible source areas by using, for example, air photos, land systems mapping and geomorphological mapping; field survey; sampling from available exposures, boreholes and trial pits; and laboratory analysis of samples.

Today, however, the geomorphologist is becoming increasingly involved in the other three areas (b), (c) and (d) described above – resource management. There is an interesting discussion of the implications of exploiting Chesil Beach in Jolliffe and McLellan (1980), and the whole topic is discussed in more detail by Cooke and Doornkamp (1974, Ch. 10).

Techniques of scenic evaluation

Although not always thought of as such, landscape is a resource just as minerals are resources. The crux of the problem of managing landscapes from an aesthetic point of view is that the conservation of landscapes for recreation, posterity or scientific study nearly always conflicts with claims from mining interests, farmers and urban developers. Furthermore, since those claims can usually be stated in numerical and economic terms, the planners face the problem of describing the aesthetic value of scenery in a precise, if not quantitative, way. There remains the associated difficulty of establishing which landscapes people enjoy, a field littered with intangibles. For these and other reasons, it is easy to criticise the various attempts that have been made to evaluate landscapes aesthetically. But decisions concerning landscape development have to be made, and the geomorphologist has a contribution to make here. This is because the features that make a landscape attractive are often of geomorphological interest, and because the geomorphologist should have an 'eye for the country'. It is true that this makes him a rather prejudiced observer, but nevertheless geomorphologists do involve themselves in this field.

Not all techniques of scenic evaluation involve a geomorphological component, but of those that do, three are especially well known. Two attempt to identify components of the landscape that contribute to one's appreciation of the scene: these are Leopold's method (1969a, b) based on site evaluation, and Linton's method (1968) based on areal classification of terrain. The third is Fines's (1968) analysis of people's personal and general responses to scenery, there being no attempt to single out individual components of the scene. A problem shared by all three is the use of quantitative techniques; landscape aesthetics does not lend itself to them. Quantification may give the impression of precision where none exists.

A subject related to this in some ways is that the layman, in addition to having an interest in scenery, also has a natural interest in landforms and their origin. A likely development over the next few years is in the area of educating the layman when he is visiting the countryside. For example, the traditional view indicator could contain a few added notes on landforms in the scene.

Land-systems mapping

This is a technique of mapping that is widely used by those involved in environmental and resource management. In fact, it was developed specifically by them and for themselves. It has many possible applications, but the main one is the mapping of information gathered during reconnaissance surveys for agricultural land and engineering projects. The area covered can be extensive, for example, a whole country or a whole desert. The area is divided up into land systems, mainly by the analysis of aerial photographs. Each land system has distinctive assemblages of surface form, materials, soils, vegetation and processes, and usually can be delimited on the basis of landform. Examples of land systems are escarpments, lowlands, plains and valleys, and thus the concept of the land system is very similar to the well known concept of the morphological region (Fig. 14.1). The

179

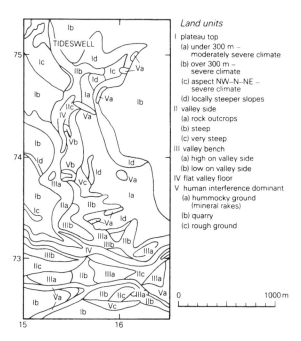

Land units

I plateau top
  (a) under 300 m –
    moderately severe climate
  (b) over 300 m –
    severe climate
  (c) aspect NW–N–NE –
    severe climate
  (d) locally steeper slopes
II valley side
  (a) rock outcrops
  (b) steep
  (c) very steep
III valley bench
  (a) high on valley side
  (b) low on valley side
IV flat valley floor
V human interference dominant
  (a) hummocky ground
    (mineral rakes)
  (b) quarry
  (c) rough ground

Figure 14.1 A land-unit subdivision of the Tideswell area, Derbyshire (from Cooke & Doornkamp 1974).

approach has been criticised as inaccurate and merely static and inventorial, but maps divided into land systems are easily understood by policy-makers and environmental managers who are not specialists in geomorphology, and the maps are also convenient ways of storing information in data banks and retrieval systems.

In this approach, the land system is regarded as the geomorphological unit, and thus contrasts interestingly with the idea considered in an earlier chapter that the drainage basin is the fundamental geomorphic unit. Land systems do not usually correspond with drainage basins. Preference for one unit or the other depends on the nature and purpose of the problem or survey. The two approaches could be combined, however, with land-systems analysis providing the general framework for systematic basin studies.

Land systems themselves can be subdivided on a map, on the basis of form, into land units. Each land unit has more or less homogeneous physical properties. Examples of land units are alluvial fan, levée, plateau top, valley-side, and group of sand dunes. Because of scale problems they would not usually be shown on a land-systems map, but would appear at the level of the site investigation.

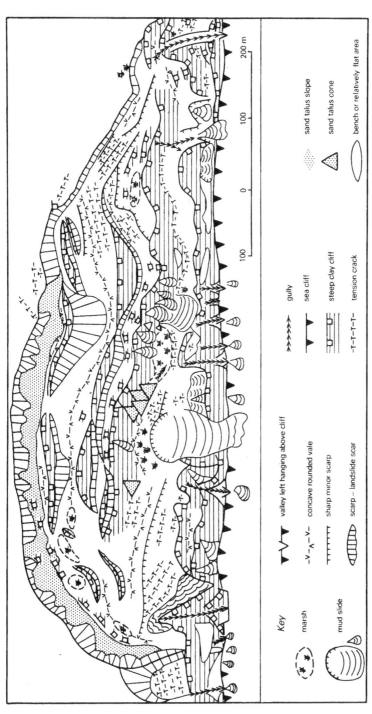

Key

▼▼	valley left hanging above cliff	⟫⟫⟫	gully
※ marsh		▼▼	sea cliff
−ᴠ−ᴧ−ᴠ−	concave rounded vale	▢▢	steep clay cliff
⬯ mud slide	┬┬┬┬	sharp minor scarp	−ᴛ−ᴛ−ᴛ−	tension crack
⟨⬯⟩		⬭	scarp – landslide scar
			sand talus slope
	△		sand talus cone
	◯		bench or relatively flat area

Figure 14.2 A geomorphological map of a coastal landslide area in Dorset (after Geological Society Engineering Group Working Party: from Cooke & Doornkamp 1974).

Geomorphological mapping

This is also a mapping technique that is of great value in applied geomorphology. It has been referred to on several occasions in the previous two chapters. The maps are of use mainly in land-use planning, hydrological engineering, civil engineering, soil surveying and conservation, but in fact the range of application is immense, as indicated in Table 14.1.

Table 14.1 Applications of geomorphological mapping in planning and economic development.

(a) Land use
territorial planning
regional area planning
conservation of the natural and cultural landscape

(b) Agriculture and forestry
potential utilisation
soil conservation
soil erosion control
reclamation of destroyed or new areas
soil reclamation
drainage and irrigation

(c) Underground and surface civil engineering
reconstruction and replanning of settlements, especially of towns
designing of industrial buildings
communications (roads, railways, canals, harbours)
hydro-engineering
   reservoirs and dams
   regulation of rivers
   natural and artificial waterways
   irrigation canals
   harbour construction
   shore protection
   fishing projects

(d) Prospecting and exploitation of mineral resources
prospecting
geological survey
exploitation
mining
potential and actual damage done by mining
reclamation of abandoned open-cast mines
landslip areas and regions of subsidence due to mining
reclamation of areas destroyed by mining and waste dumps

Source: Demek (ed.) (1972) *Manual of detailed geomorphological mapping*. Academia.

An example of a geomorphological map is given in Fig. 14.2. Each landform is given a coloured symbol, and the symbols are then marked on the map in the correct places. It is also possible to map surface materials and geomorphological processes in the same way. An extensive, although by no means exhaustive, list of symbols is given in Fig. 14.3. A geomorphological map could be produced at more or less any scale, but because its greatest value lies at the level of site investigation it would usually have a scale between 1 : 10 000 and 1 : 75 000. It is, therefore, compatible with land-unit mapping rather than land systems.

The technique dates from the early 1950s, and its widespread use reflects the fact that a valuable aid in many environmental management problems is a map of the relevant landforms, materials and processes. In terms of their practical value, the maps fall into two categories:

(a) *The general geomorphological map.* This is simply a means by which the geomorphological characteristics of an area may be recorded in map form. It has some value at the initial field investigation stage, but being neither selective nor related to any particular problem it is of limited practical application.

(b) *The purpose-orientated map.* This carries information relevant only to a particular investigation, and can be presented either as just a part of a general map or as a product of the use of the general map as a base map to which specific detail can be added.

Maps showing just surface form – detail about slope angles and breaks of slope – are called morphological maps (Fig. 14.4). Maps can also show just materials – solid rock and superficial deposits – and this might be of use in an engineering problem. Maps rarely show only processes, but process can be inferred from form. For example, the appearance of a gully, recorded on a geomorphological map, might be the first sign of soil erosion.

Conclusion

To show the versatility of the mapping techniques just discussed, this chapter concludes with an unusual application of terrain analysis: the military application. This also shows that the term 'resource' must be viewed as having a wide meaning: landscape has been considered as a materials resource and an aesthetic resource, and now, here, it is seen as a strategic resource.

Terrain has always been a factor in military operations, but the specific application of professional geomorphological investigations is

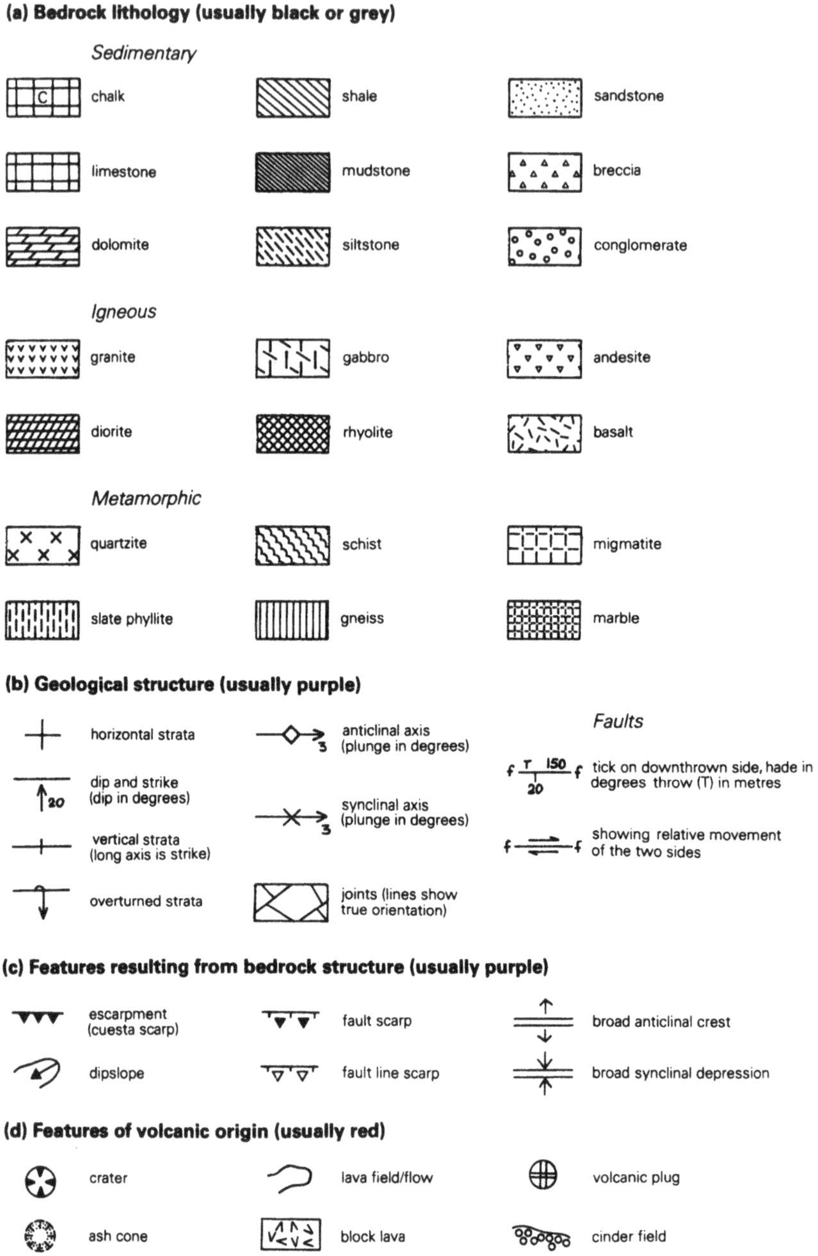

(a) Bedrock lithology (usually black or grey)

Sedimentary

[symbol]	chalk	[symbol]	shale	[symbol]	sandstone
[symbol]	limestone	[symbol]	mudstone	[symbol]	breccia
[symbol]	dolomite	[symbol]	siltstone	[symbol]	conglomerate

Igneous

[symbol]	granite	[symbol]	gabbro	[symbol]	andesite
[symbol]	diorite	[symbol]	rhyolite	[symbol]	basalt

Metamorphic

[symbol]	quartzite	[symbol]	schist	[symbol]	migmatite
[symbol]	slate phyllite	[symbol]	gneiss	[symbol]	marble

(b) Geological structure (usually purple)

+	horizontal strata
—	dip and strike (dip in degrees)
+20	
+	vertical strata (long axis is strike)
⊤	overturned strata

anticlinal axis (plunge in degrees)

synclinal axis (plunge in degrees)

joints (lines show true orientation)

Faults

f ⊤ 150 f   tick on downthrown side, hade in
   ——   degrees throw (T) in metres
   20

f ⇌ f   showing relative movement of the two sides

(c) Features resulting from bedrock structure (usually purple)

▼▼▼	escarpment (cuesta scarp)	▼▼▼	fault scarp	↑	broad anticlinal crest
[symbol]	dipslope	▽▽	fault line scarp	↓↑	broad synclinal depression

(d) Features of volcanic origin (usually red)

[symbol]	crater	[symbol]	lava field/flow	[symbol]	volcanic plug
[symbol]	ash cone	[symbol]	block lava	[symbol]	cinder field

Figure 14.3 Geomorphological mapping symbols: in (a) and (d), the shading patterns have been chosen from commercially available systems (from Cooke & Doornkamp 1974).

cinder cone

 ropy lava

 dike

 caldera

 pillow lava

 geyser

(e) Superficial unconsolidated materials (usually black or grey)

 duricrust

 gravel

 clay

 boulders angular

 sand

 shells

 boulders rounded

 silt

 peat

(f) Slope instability features (usually brown)

 landslide (type undetermined)

 flow slide

 rock fall

 rotational slide (with backtilt)

 mud slide

 solifluction lobe

 non-circular rotational slide with graben

 sand run

 soil creep

(g) Aeolian features (usually yellow)

 windblown sand plain

 parabolic dune

 deflation area

 plain with desert movement

 longitudinal dune

 deflation hollow

 foredune (marginal dune)

 barchan

 sand mounds around vegetation

(h) Coastal features (usually green)

 surf zone

 cliff face

 offshore bar

 prevailing drift

 abandoned cliff face

 spit

 rock shore platform

 valleys left hanging by cliff recession

 beach ridges

 (valley cross section shown)

 lagoon

 fan

(i) Forms of permafrost areas, glacial and periglacial features (usually light blue)

snowfield

drumlin

dead ice depression

glacier ice

roches moutonnées

glacial outwash

ground moraine

cirque

thaw basin

rock bar

terminal moraine

thaw subsidence

glacial trough

lateral moraine

pingo

hanging glacial valley

medial moraine

patterned ground

avalanche track

esker

stone stripes

kame deposits

(j) Forms of fluvial origin (usually dark blue)

stream		levées		permanent lake	
river channel		point bar		temporary lake	
dry river channel		erosion terrace		area susceptible to flooding	
waterfall (rapids)		accumulation terrace		fluvial erosion on slopes	
spring		alluvial fan		rill	
sand bar		delta		gully	
cut-off meander		swamp		sheet	
oxbow lake				badlands	

(k) Karst landscape features (usually orange)

conical karst		limestone pavement		swallow hole	
tower karst		clints and grikes		cave	
labyrinth karst		doline		gorge	

(l) Major features not included in previous figures (usually brown)

planation surface		rock wall		pediment	
residual hill		pass (col)		ridge	

(m) Man-made features (usually black or grey)

quarry		mining pits		embankment	
sand pit		tips (mounds)		breakwater	
gravel pit		filled hollows		settlement area	
mining subsidence		transport route		surface heavily remodelled	

Figure 14.3 *continued*

fairly recent. Military tactics involve landforms and geomorphic processes in five ways.

(1) The disposition of high and low ground for vantage and refuge.
(2) The suitability of terrain for cross-country movement.
(3) The suitability of terrain for establishing permanent or semi-permanent military positions such as trenches and camps.
(4) The availability of raw materials, especially sand and gravel.
(5) Natural hazards, such as floods.

In a recent article, Mitchell and Gavish (1980) discuss a number of interesting aspects of this subject. They show how land-systems mapping can be used to predict terrain conditions in virtually 'unknown' areas where the only information is a topographic map, a geological map and some air photos; they discuss the possibility of

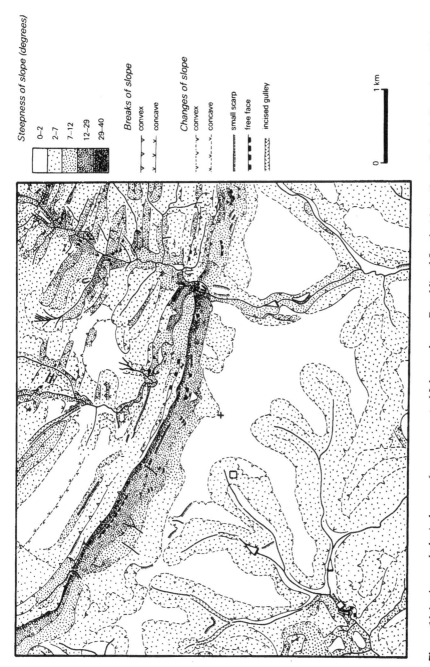

Steepness of slope (degrees)

0–2
2–7
7–12
12–29
29–40

Breaks of slope

convex
concave

Changes of slope

convex
concave

small scarp
free face
incised gulley

0   1 km

Figure 14.4 A morphological map of an area west of Johannesburg, Republic of South Africa (from Cooke & Doornkamp 1974).

keeping data banks of the results of military terrain analyses; and they illustrate the general principles of terrain analysis with examples from Israel–Egypt conflicts in the Sinai peninsula. They also show that future terrain prediction techniques will almost certainly be based on remote-sensing imagery.

PART V

The current conceptual status of geomorphology

15 Pure and applied geomorphology in context

Introduction

This final chapter aims to put developments in both applied and pure geomorphology into context. The rise of applied geomorphology has cast the professional geomorphologist in the role of decision-maker, or at least a member of a decision-making team. Moreover, these decisions usually have financial implications, so that his work can become the subject of cost–benefit analysis. However, some of the most important and interesting decisions may involve social or aesthetic factors that cannot be expressed in financial terms, so that the decision-making process becomes more complicated. Applied studies also increase the amount of contact between geomorphology and other disciplines. The contact between pure and applied geomorphology is also considerable, as results from the 'real world' are fed back into pure research so that refined theory can then be applied in future to practical problems.

The history of geomorphology is reviewed in terms of paradigm shifts, and the current status of the subject is discussed.

The position of applied geomorphology

Where the effective control lies

It is beyond dispute that, to be succesful, planners, environmental managers and developers need to be well informed about the nature of geomorphological problems, and therefore need the services of the professional applied geomorphologist. However, since in studying applied geomorphology we are analysing the relevance of geomorphology to the general needs of society, it is pertinent to establish exactly what the role of the geomorphologist is in the decision-making process. Where does the control lie?

The answer is that effective control lies with an enormous number of individuals, not all of whom are trained environmental scientists. Therefore, the crucial aspect of environmental management is to bring the scientist and the environmental decision-maker together. There is also an element of public consultation, including the farmer, forester and tourist.

Many private firms and government agencies exist which deal with some aspect of the Earth's geomorphological systems. From this it follows that geomorphological information is required by all kinds of organisations concerned with developing or managing the environment. At an international level, UNESCO, in its concern for the welfare of mankind, has acknowledged the important role of applied geomorphology. At the national level there are a number of government bodies, such as national planning units and departments of hydrology and agriculture, which require such information. Regional and local planners also make use of geomorphological contributions. Private firms include consulting engineers, land developers, land managers, and those dealing with the legal aspects of planning. Table 11.1 (p. 134) indicates the range of contributions that geomorphology can make in various types of planning. The applied geomorphologist usually is not working alone, and the final decision will probably be made by someone else. He generally works as a member of a surveying or planning team, and the information he provides goes into a data bank derived from a range of disciplines. He frequently works alongside other specialists including civil engineers, agriculturalists, economists, geologists, geotechnical experts, soil scientists and agricultural botanists.

The interdisciplinary character of applied geomorphology will be discussed shortly, and in the next section we discover that the geomorphologist now commonly acts as one of a team of consultants for a decision-making body such as a government department or a private agency. He therefore becomes an important member of the professional decision-making community. It is a new and honourable role for him to play, but it does have the effect of drawing geomorphological work into the open. Geomorphology is no longer a purely academic subject where ideas and theories can be tossed around by intellectuals. The subject, its theories and conclusions have to 'work' – they have to help to solve real problems. It demands an accuracy, precision and scientific rigour which has not been required before. In short, it is a challenge. Other sciences, such as medicine, are expected and thought to be entirely credible. Geomorphology now finds itself in the same position.

The socio-economic importance of decision-making

Having established that there is a decision-making process in which the applied geomorphologist plays a part, it is interesting to go on to ask: What is the economic value of such decisions? It is clear that geomorphology, and physical geography in general, long seen as of purely scientific interest, is now of considerable economic importance. This is not confined just to assessing the value of individual resources such as sand and gravel. Probably even more important in the long run is the application of knowledge about the nature of materials and the nature of erosional and depositional processes to matters such as natural hazards, construction sites and the sensible management of the environment. There can be a vast element of financial loss in flood damage, soil erosion and slope failure.

In economic terms, the decision usually reduces to a considerable short-term expenditure having to be weighed against long-term benefits. Sometimes alternative solutions to a problem have to be compared one against the other. The decision can then become extremely complex, because environmental problems invariably involve so many factors. In systems terms, what is happening is that a socio-economic system is intersecting with a geomorphic process–response system to produce a geographical control system. The situation lends itself to the use of computers programmed on a systems basis to carry out a cost–benefit analysis. The decision is then a purely economic one, based on comparing the benefits that accrue with the price that has to be paid. Two examples of such analyses from the field of water resources are discussed by Chorley and Kennedy (1971, pp. 333–41).

However, a final problem with decision-making in environmental management is that the problem rarely is a purely economic one; it is socio-economic. In a given scheme, the social costs and the social benefits have to be included in the cost–benefit equation. But often they cannot be. They cannot be expressed satisfactorily just in terms of money. Subjective assessments are getting mixed up with objective assessments. The same problem recurs when environmental quality, such as some impressive scenery, is involved as well. The value of scenery cannot be measured in pounds. In environmental management, it is a difficulty that has yet to be resolved satisfactorily.

To a reader not familiar with this sort of line of argument some of this may seem rather abstract. So let us look at an example where these issues are involved. In England a long-standing idea is that a possible addition to the motorway network is a link between Manchester and Sheffield via Longdendale (Fig. 15.1). The long-term value of the motorway can be measured, albeit not particularly accurately, in economic terms. So can the cost of constructing the motorway, given a decision on the route and certain details about it. This is an exercise in cost–benefit analysis, deciding whether the motorway should be built and choosing between possible alternative routes. The applied geomorphologist will already have made a contribution, supplying

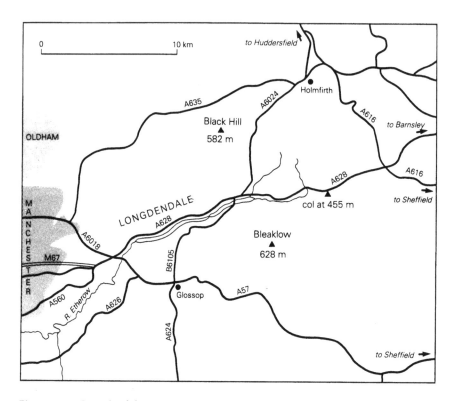

Figure 15.1 Longdendale.

information about surface materials, slope stability, drainage problems and so on.

But it is not as simple as that. The motorway, if built, might disrupt farmland, farmers, the lives of people, even whole villages. To some extent this can be measured in economic terms: financial compensation exists. But not completely, for the full social cost of disruption cannot be measured in money.

There is one further complication. The route of the motorway through Longdendale would lie partly in the Peak District National Park. A host of conservation issues now enter the discussion: preservation of wilderness areas; preservation of beautiful scenery; impact on wildlife; noise, pollution, and so on. Against this, Longdendale has already been despoiled by the construction of reservoirs, railway lines and power lines (Fig. 15.2) so in some ways this would be a suitable route for a motorway. Also, the motorway would give access to the Peak District, and one of the purposes of the National Park idea is to make them accessible. Whether or not the applied geomorphologist is in a position to advise on this area of

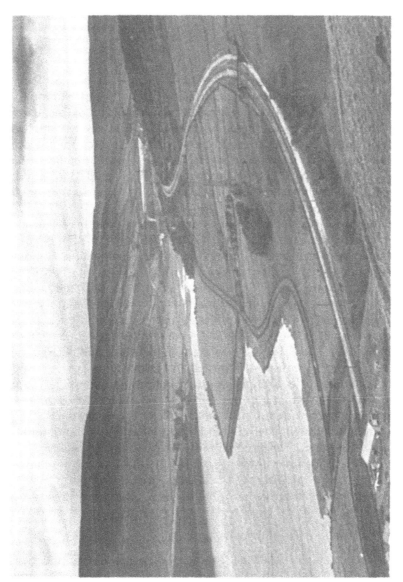

Figure 15.2 A view of Longdendale showing existing despoliation (from Walker 1977).

debate is not the point. The point is that the decision has now been made more difficult by the appearance of important issues which cannot be valued in purely financial terms. It is often thus. As it happens, at the time of writing, the idea of building the motorway has been shelved.

Links with other disciplines

Geomorphology has always proved very difficult to define. This is partly because it has extensive areas of overlap with other disciplines which blur its identity as a subject. Geomorphology makes use of information and ideas from several other subjects, such as biology, physics, chemistry, geology, geography, botany, oceanography, glaciology, pedology, hydrology and hydraulics. From this it follows that almost the entire field of geomorphology is covered one way or another by other subjects. It gives many geomorphologists a feeling of unease and lack of identity.

The rise of applied geomorphology at once reinforces this reliance on other subjects yet at the same time makes it unimportant. Problems in environmental management require an understanding of a complete geomorphic system, and this may include influences that lie outside traditional geomorphology. One effect of this is that the applied geomorphologist is now brought into contact with even more subject areas than before, such as economics, planning, law and psychology. Another effect is that the applied geomorphologist finds himself working with a variety of environmental managers and technicians: engineers, farmers, foresters, planners and politicians.

The boundary line around geomorphology is, therefore, more blurred than ever. Yet, in a sense, that does not matter. Environmental management recognises no traditional subject divisions; it is a truly interdisciplinary study. The most important issue is not the definition of geomorphology but the environmental problem under consideration. That problem needs everyone's full attention and energy. Now that we are dealing with important and even serious practical problems, parochial concerns about geomorphology as a subject are irrelevant.

The feedback to pure geomorphology

It has been established that geomorphology is concerned with landforms, materials and their related processes. In practice, its study now falls into two categories, modern pure geomorphology and modern applied geomorphology. The two exist side by side, as in mathematics. They are recognisably distinct disciplines, yet they rely one upon the other. The massive development of applied geomorphology during the past 10 years has not meant that pure geomorphology is now less important. Indeed, every advance in the applied side brings a corresponding advance in the pure side. This is because the applied branch of the subject relies on the pure branch for information, knowledge, principles, theories and laws that can be applied to practical problems. Then, as Verstappen foresaw in 1968, applied geomorphology produces hard facts about what happened in practice that can be fed back into pure research. The process is then repeated and refined. Pure geomorphology supplies the demand from applied geomorphology. Their relationship is the same as that between science and technology. Technology is the practical application of scientific advances. Pure geomorphology is a science. Applied geomorphology is technology. It sets geomorphology firmly in the family of sciences and gives the subject a unity of purpose that it has probably never had before.

Approaches to geomorphology

Paradigms in science

The role of paradigms in the progress of science has been analysed by Kuhn (1962). It must be stressed at the outset that Kuhn was writing about science in general, not about geomorphology in particular, and in fact he drew most of his examples from physics and chemistry. His basic idea was to challenge the traditional view that the development of science takes place by the gradual accumulation of discoveries and inventions. Instead, he suggested that progress occurs in a more irregular way, with revolutions occurring by the replacement of one paradigm by another. Therefore, the origin, continuance and obsolescence of paradigms together occupy an important place in the history of evolution of science.

The term 'paradigm' is not easy to define. It is an accepted pattern of beliefs that provides the framework for the study of a particular science. It is an assemblage of ideas and concepts that underpin a science at any particular time. It is an approach to science, a way of looking at things. It is a stable pattern of scientific activity. In a sense it is a model for intellectual thought and research, but it is more than a model: it is a sort of 'super-model' within which other models are set.

It is a legitimate question to ask why paradigms are necessary, or, to put it more elegantly, to ask why and if scientific thinking is characterised by the existence of paradigms. And the answer lies in another question: What constitutes a relevant fact? In other words, in a science, say geomorphology, out of all the innumerable facts that we could collect, how do we know which ones are relevant? And the answer is that we don't, unless we have some kind of general overview – paradigm – that defines fields of study that are relevant and those that are not. It follows that, in the absence of a paradigm, fact-gathering is random and, in these circumstances, the mere accumulation of data can produce a veritable morass. To that extent, the acquisition of a paradigm is a sign of maturity in the development of a science.

In a given science, when a particular paradigm has outlived its usefulness it is, or can be, replaced by another one. This is called a 'paradigm shift', and it constitutes a revolution in that science. This is reflected in Hallam's (1973) selection of a title for his book about plate tectonics, *A revolution in the earth sciences*. Kuhn showed that during the period following the wholesale adoption of a new paradigm, research workers are confronted by three problems. The first is that of acquiring a body of significant knowledge. A curious effect of paradigm adoption is that a period of abundance of data is replaced by one of scarcity. It takes time to build up a stock of data relevant to the new paradigm. The second is the matching of facts, old and new, to new theory. This provides the focus for research activity. The paradigm allows the progressive development of models within its general terms. The third is that the range of acceptable work suddenly narrows. The internal discipline of the paradigm, its unwritten rules and traditions, guides the pattern of research along new avenues. Paradigms are restrictive. They concentrate work in particular fields but in doing so they give the science unity and purpose.

For an ingenious idea to become a paradigm, it needs to have certain characteristics, certain minimum ingredients for success. Kuhn identifies three:

(1) It must be able to solve at least some of the problems that have brought the old paradigm to crisis point.
(2) It must appeal to the academics' sense of what is elegant, appropriate and simple.
(3) It must contain more potential for expansion than the old.

In the following section, and to conclude this book, I propose to discuss paradigms and paradigm shifts with reference to geomorphology. This should help the reader to make sense of the difficult ideas that have just been introduced because he will be able to relate them to a familiar subject. But that is not the real reason for ending the book in this way. The real reason is that, thinking back to my own schooldays and early undergraduate days in the 1960s, geography (including geomorphology) was just a collection of information. I had no idea that there was some kind of logical thinking that had continued for 100 years or more that lay behind the selection of that information for my consumption, nor that there was a revolution in that thinking going on at that very time. This constitutes no criticism of my own teachers. The point is that in some subjects, such as English literature and history, knowledge of the nature of intellectual debate in that subject is an important aspect of scholarship. I think it should be in geomorphology, too. The major landmarks in the subject are tabulated in Table 15.1.

Paradigm shifts in geomorphology

Discussion of this subject has two problems. First, Kuhn wrote his important book in 1962, so it follows that paradigms and paradigm shifts that took place before then were not recognised as such at the time. They were seen as paradigms in retrospect. Secondly, some of the relevant paradigms are, or were, not strictly geomorphological, since some of them were paradigms in geology or geography, so that they became paradigms in geomorphology by its association with those two subjects.

It might be thought that the first paradigm in geomorphology was Davis's cycle of erosion, but in fact there is no reason why we cannot go back much further than that and start with catastrophism. Belief in the Creation and the Flood, or some other catastrophist idea, lay at the core of geological thought for hundreds of years. The sureness with which these beliefs were held can be judged from the difficulty with which they were displaced. The only sense in which catastrophism cannot be described as a paradigm is that it was less of a framework and more a body of 'facts' that turned out to be wrong. As described in Chapter 1, one of the factors leading to the eventual demise of catastrophism was acceptance of the Glacial Theory. This theory in itself does not amount to a paradigm, although much geological and geomorphological thought does depend on it. Actually, of course, European Pleistocene glaciation is not a fact either, merely a highly probable inference.

It was stated earlier that the acquisition of a paradigm is a sign of maturity in the development of a science, and in the early 19th century the twin paradigms of uniformitarianism and stratigraphical correlation by fossils established geology as a true science. The idea that uniformitarianism is an early paradigm in geomorphology is almost without question, and yet in a way 'paradigm' is the wrong term for uniformitarianism, because the concepts it embodies are too deep, too basic for that. It is a truth that can never be displaced by a paradigm shift. In much the same way, the Law of Gravitation is not the current paradigm in physics; it underpins the subject too deeply for that. Perhaps science needs a hierarchy of paradigms. If paradigms are super-models, then the subject needs a term for a 'super-paradigm' to

Table 15.1 The major landmarks in geomorphology.

1788	Hutton's *Theory of the Earth*
1802	Playfair's *Illustrations of the Huttonian theory of the Earth*
1815	William Smith: superposition, and correlation by fossils
1830	Lyell's *Principles of geology*
1840	Agassiz: the Glacial Theory
1859	Darwin's *On the origin of species*: evolution
1875	Powell's *Exploration of the Colorado River of the West and its tributaries*
1877	Gilbert's *Report on the geology of the Henry Mountains*
mid-1880s	term 'geomorphology' introduced
1888	Suess: the concept of eustasy
1899	Davis proposes the erosion cycle
1909	Davis's *Geographical essays*
	Penck and Brückner model: four glacials
1912	Wegener's continental drift
1914	Gilbert's *Transportation of debris by running water*
1924	Walter Penck's morphological analysis of landforms (in German)
1928	Baulig: the origins of denudation chronology
1938	W. M. Davis's last publication
1941	Bagnold's *Physics of blown sand and desert dunes*
1945	Horton: drainage basin morphometry and the origin of quantification (?)
1950	Strahler: early application of statistics
	L. C. King's pediplanation theory
1953	Penck's *Morphological analysis of landforms* (English translation)
1955	Wooldridge and Linton: denudation chronology reaches its peak
1960	approximate date for the end of the erosion cycle and denudation chronology
1962	Kuhn introduces the term 'paradigm'
	Chorley: geomorphology and systems theory
	early use of term 'plate tectonics'
1964	Leopold, Wolman and Miller
1965	a large number of statistical concepts introduced between 1960 and 1965
1966	Manley writes of the 'new geography'
1967	Chorley and Haggett's *Models in geography*
1969	Chorley's *Water, earth and man*
1970	applied studies begin to multiply noticeably
	Craik's 'Environmental psychology', on perception
1971	Chorley and Kennedy's *Physical geography: a systems approach*
	advanced statistical techniques introduced
1972	the 'new geology' (plate tectonics) well established
1974	Cooke and Doornkamp's synthesis of applied studies in geomorphology
1976	Whalley's *Properties of materials and geomorphological explanation*
1980	Coates and Vitek's *Thresholds in geomorphology*
1980–83	man–environment studies come to the fore

The rows for 1875 and 1877 are bracketed together and labelled: early process geomorphology encompass very basic tenets like uniformitarianism.

When Davis's erosion cycle appeared, therefore, it made use of uniformitarianism, it did not replace it. There was no paradigm shift. In this sense, the erosion cycle is the first paradigm in geomorphology, in that ultimately it was replaceable. As described in earlier chapters, the cycle of erosion remained the paradigm, at least in Britain and the United States, for about 50 years.

The identification of denudation chronology as a paradigm raises some interesting points. It is totally justified in that it has all the characteristics and properties required by Kuhn. And yet it was born directly out of the erosion cycle, so that the two were always very close together. One was a specific application of the other. Again, there was no paradigm shift as denudation chronology came in. Later, when one died, they both died. As described earlier, it is difficult to put a date on this event, but 1960 is a convenient round figure.

In view of Kuhn's idea about the immediate replacement of one paradigm by another, something which must be rather unusual in science happened in geomorphology in the early 1960s. The cycle of erosion and denudation chronology were not replaced for several years. There was a conceptual vacuum until about 1967, an inter-regnum. It meant that geomorphologists were consciously searching around for a paradigm with which to replace them. Some would say that they still are. Certainly the period since then has been one of frantic activity in the search for a paradigm, and the speed of events has perhaps revealed geomorphology as a rather immature science.

The first entry into the field was the model-based paradigm. At the time when Chorley and Haggett's influential book appeared, in 1967, geography inherited a paradigm that was largely classificatory. In Chapter 1 of that book they argue that a model-based paradigm could prove to be a viable and forward-looking alternative for geography in general. In Chapter 3 Chorley puts the case specifically for geomorphology, although by implication rather than by directive. Now the useful simplification of reality, that is, model-building, plays an important part in our attempt to solve geomorphological problems, and yet if a number of geomorphologists were asked to name the current paradigm in their subject, few, if any, would say models. That is partly because of the current disenchantment with models among some geographers, but also because, back in the 1960s, before we could see if models would form the basis of a new paradigm, another development came along which overtook it. This was General Systems Theory.

There is a close relationship between models and General Systems Theory, confusingly so to some people. Chorley's chapter on geomorphology in *Models in geography* (1967) in fact deals almost entirely with systems. Systems, or at least the way we express them on paper, are also models. More to the point, however, almost all models can be included in systems analysis, so that General Systems Theory includes models and at the same time widens the field still further. Chorley and Kennedy's (1971) book on systems analysis in geomorphology does not actually mention the word 'paradigm' anywhere, but nevertheless the book is heavy with implication that systems analysis does offer a new paradigm for the subject. Although during the 15 years since then the response to systems analysis has fallen short of complete aceptance, there is much evidence that systems thinking does form the current paradigm in geomorphology. Many textbooks use the systems approach, and virtually any aspect of modern geomorphology – whether process, form, materials, man, applied – lends itself, and profitably so, to systems analysis. In doing so, geomorphologists are adopting a paradigm that is very basic indeed. General Systems Theory, in a sense, is the paradigm for all science (see, for example, Boulding 1956). From a geomorphological point of view that is both an advantage and a disadvantage. The adoption of the systems paradigm sets the subject firmly in amongst the other sciences, and yet it is not distinctively geomorphological. One thing is very clear. Systems theory is a true paradigm, being an approach to the study of science, not a body of knowledge in itself. To that extent it can never be shown to be 'wrong'. Any future paradigm shift will come only when the systems way of looking at things is no longer useful.

All that sounds very neat and conclusive, but that is not, in fact, the end because there have been three other developments since the early 1970s that are relevant to this discussion. The first is the adoption of plate tectonics as a paradigm by the earth sciences. It is difficult to put a date to this since plate tectonics grew over a number of years out of the ideas of continental drift and sea-floor spreading, but by 1973 Hallam was able to write of a revolution in the earth sciences and that 'in Earth Sciences, it is quite clear that Plate Tectonics is the currently held paradigm'. The whole book is a very interesting read on what it takes to achieve a consensus among working scientists. Hallam attempts to identify the existing paradigm that it replaced, and he also goes on to make the interesting point that, within the broad field of earth science, a number of subjects such as sedimentology and geochemistry had made rapid progress recently without reference to plate tectonics. Although he did not do so, he could have included geomorphology, or at least most of it, in that list. The implications of plate tectonics for geomorphology were discussed earlier, and it is clear that plate tectonics has a bearing on a number of topics that geomorphologists have traditionally studied: mountain-building, folding, faulting, rift valleys, island arcs, volcanoes, earthquakes, the major structural features of the Earth. To that extent, plate tectonics cannot be ignored by geomorphologists, and yet it can never be a paradigm for geomorphology because it impinges on only a very small part of the subject. Perhaps it shows that geomorphology is not really one of the earth sciences, or that only part of it is.

The second important development is thresholds. This was put forward as a conceptual basis for all geomorphology by Coates and Vitek (1980). However, although they demonstrate the wide application of the threshold concept in geomorphology, it seems unlikely that, in itself, it could form a paradigm for the whole subject. The threshold is merely one element in systems analysis, so that the concept is already catered for by having General Systems Theory as the paradigm.

The third development is the renewed focus on man–land inter-relationships. It seems ironic that this 'new' idea has, in various guises, always been at the centre of philosophical discussion within geography. Some workers now view this as the main theme in geomorphology and geography, but even if it is it seems unlikely to constitute a paradigm in itself. Again, the whole idea is contained within General Systems Theory: control systems, and man's intervention within natural systems. It will be interesting to see if these new developments reunite geography as a coherent subject, as they promise to do.

Conclusion

The history of geomorphological thought has been characterised by a number of paradigm shifts, as one intellectual proposition replaced another as a framework for the subject. Until the mid-1960s, these changes were restricted to pure geomorphology. The appearance at that time of the model-based approach, however, had the effect of strengthening the ties between geomorphology and the rest of geography, since models were applicable to geography in general. This trend has continued with the emergence of man–land interrelationships, hazards and applied studies as new, or renewed, themes in geomorphology, as all three explore the line of contact between geomorphology and human geography. At the time of writing, therefore, geomorphology stands as an important applied science in its own right but with strong links with its parent subject, geography.

Bibliography

Agassiz, L. 1840a. *Études sur les glaciers*. Neuchâtel.

Agassiz, L. 1840b. On glaciers, and the evidence of their having once existed in Scotland, Ireland and England. *Proc. Geol Soc. Lond.* **3**, 327–32.

Allen, J. R. L. 1970. *Physical processes of sedimentation*. London: George Allen & Unwin.

Andrews, J. T. 1971. *Techniques of till fabric analysis*. GeoAbstracts Technical Bulletin, no. 6. Norwich: GeoAbstracts.

Andrews J. T. 1975. *Glacial systems*. Duxbury.

Bagnold, R. A. 1941. *The physics of blown sand and desert dunes*. London: Methuen.

Balchin, W. G. V. 1952. The erosion surfaces of Exmoor and adjacent areas. *Geogr. J.* **118**, 453–76.

Baulig, H. 1928. *Le plateau central de la France*. Paris: Colin.

Baulig, H. 1935. *The changing sea-level*. Inst. Br. Geogs Publ., no. 3.

Baulig, H. 1938. Questions de terminologie. *J. Géomorph.* **1**, 224–9.

Baulig, H. 1940. Le profil d'équilibre des versants. *Ann. Géogr.* **49**, 81–97.

Baulig, H. 1950. *Essais de géomorphologie*. Paris: Les Belles Lettres.

Baulig, H. 1952. Surfaces d'aplanissement. *Ann. Géogr.* **61**, 161–83, 245–62.

Bennett, R. J. and R. J. Chorley 1978. *Environmental systems*. London: Methuen.

Bertalanffy, L. von 1956. General system theory. *General Systems Yearbook* **1**, 1–10.

Bird, E.C.F. 1968. *Coasts*. Canberra: Australian National University Press.

Birkeland, P. W. 1974. *Pedology, weathering and geomorphological research*. Oxford: Oxford University Press.

Blackwelder, E. 1933. The insolation hypothesis of rock weathering. *Am. J. Sci.* **26**, 97–113.

Bloom, A. L. 1969. *The surface of the earth*. Englewood Cliffs, NJ: Prentice-Hall.

Bloom, A. L. 1978. *Geomorphology: a systematic analysis of Late Cenozoic landforms*. Englewood Cliffs, NJ: Prentice-Hall.

Boer, G. de 1977. Coastal erosion. In *The unquiet landscape*, D. Brunsden and D. C. Doornkamp (eds), 73–8. Newton Abbot: David & Charles.

Bögli, A. 1980. *Karst hydrology and physical speleology*. Berlin: Springer-Verlag.

Bolt, B. A. 1978. *Earthquakes – a primer*. San Francisco: W. H. Freeman.

Bolt, B. A., W. L. Horn, G. A. Macdonald and R. F. Scott 1975. *Geological hazards*. New York: Springer-Verlag.

Boulding, K. 1956. General systems theory – the skeleton of science. *General Systems Yearbook* **1**, 11–17.

Bowen, D. Q. 1978. *Quaternary geology: a stratigraphic framework for multidisciplinary work*. Oxford: Pergamon Press.

Bretz, J. H. 1962. Dynamic equilibrium and the Ozark land forms. *Am. J. Sci.* **260**, 427–38.

Briggs, K. 1977. *Sediments*. London: Butterworth.

Brown, E. H. 1957. The physique of Wales. *Geogr. J.* **123**, 208–30.

Brown, E. H. 1960. *The relief and drainage of Wales*. Cardiff: University of Wales Press.

Brown, E. H. 1961. Britain and Appalachia: a study in the correlation and dating of planation surfaces. *Trans Inst. Br. Geogs* **29**, 91–100.

Brown, E. H. 1970. Man shapes the earth. *Geogr. J.* **136**, 74–84.

Brunsden, D. (ed.) 1971. *Slopes, form and process*. Inst. Br. Geogs Spec. Publ., no. 3.

Brunsden, D. 1977. Landslides. In *The unquiet landscape*, D. Brunsden and J. C. Doornkamp (eds), 37–42. Newton Abbot: David & Charles.

Brunsden, D. 1981. Trail blazers for highway engineers. *Geogr. Mag.* **53**, 531–3.

Brunsden, D., J. C. Doornkamp and D. K. C. Jones 1978. Applied geomorphology: a British view. In *Geomorphology: present problems and future prospects*, C. Embleton, D. Brunsden and D. K. C. Jones (eds), Ch. 15. Oxford: Oxford University Press.

Bryan, K. 1925. *The Papago country, Arizona.* US Geol. Surv. Wat. Supp. Pap., no. 499.

Buckland, W. 1823. *Reliquiae diluvianae.* London: John Murray.

Büdel, J. 1963. Klima-genetische geomorphologie. *Geog. Rundschau* **7**, 269–86.

Büdel, J. 1969. Das system der klima-genetischen geomorphologie. *Erdkunde* **23**, 165–82.

Büdel, J. 1982. *Climatic geomorphology.* Princeton, NJ: Princeton University Press.

Bury, H. 1910. On the denudation of the western end of the Weald. *Q. J. Geol Soc.* **66**, 640–92.

Calder, N. 1972. *Restless earth: a report on the New Geology.* London: BBC Publications.

Carroll, D. 1970. *Rock weathering.* New York: Plenum Press.

Carson, M. A. 1971. *The mechanics of erosion.* London: Pion.

Carson, M. A. and M. J. Kirkby 1972. *Hillslope form and process.* Cambridge: Cambridge University Press.

Charpentier, J. de 1835. Notice sur la cause probable du transport des blocs erratiques de la Suisse. *Annls. Mines, Paris* **3**, 8, 219–36.

Chorley, R. J. 1962. *Geomorphology and general systems theory.* US Geol. Surv. Prof. Pap., no. 500B.

Chorley, R. J. 1965. A re-evaluation of the geomorphic system of W. M. Davis. In *Frontiers in geographical teaching*, R. J. Chorley and P. Haggett (eds), Ch. 2. London: Methuen.

Chorley, R. J. 1966. The application of statistical methods to geomorphology. In *Essays in geomorphology*, G. H. Dury (ed.), 275–387. London: Heinemann.

Chorley, R. J. 1967. Models in geomorphology. In *Models in geography*, R. J. Chorley and P. Haggett (eds), Ch. 3. London: Methuen.

Chorley, R. J. (ed.) 1969a. *Water, earth and man.* London: Methuen.

Chorley, R. J. 1969b. The drainage basin as the fundamental geomorphic unit. In *Water, earth and man*, R. J. Chorley (ed.). London: Methuen.

Chorley, R. J. 1969c. *Introduction to fluvial processes.* London: Methuen.

Chorley, R. J. (ed.) 1972. *Spatial analysis in geomorphology.* London: Methuen.

Chorley, R. J., A. J. Dunn and R. P. Beckinsale 1964. *The history of the study of landforms or the development of geomorphology*, Vol. 1: *Geomorphology before Davis*. London: Methuen.

Chorley, R. J., A. J. Dunn and R. P. Beckinsale 1973. *The history of the study of landforms*, Vol. 2. *The life and work of William Morris Davis*. London: Methuen.

Chorley, R. J. and P. Haggett (eds) 1967a. *Models in geography*. London: Methuen. (Later published in two volumes, 1969.)

Chorley, R. J. and P. Haggett 1967b. Models, paradigms and the New Geography. In *Models in geography*, R. J. Chorley and P. Haggett (eds), Ch. 1. London: Methuen.

Chorley, R. J. and B. A. Kennedy 1971. *Physical geography: a systems approach.* London: Prentice-Hall.

Chorley, R. J., S. A. Schumm and D. E. Sugden 1985. *Geomorphology.* London: Methuen.

Clayton, K. M. 1953. The glacial chronology of part of the middle Trent basin. *Proc. Geol Assoc.* **64**, 198–207.

Clayton, K. M. 1971. Reality in conservation. *Geogr. Mag.* **44**, 83–4.

Coates, D. R. 1980. Evidence for lawyers. *Geogr. Mag.* **53**, 48–9.

Coates, D. R. (ed.) 1981a. *Coastal geomorphology.* London: George Allen & Unwin.

Coates, D. R. (ed.) 1981b. *Geomorphology and engineering.* London: George Allen & Unwin.

Coates, D. R. and J. D. Vitek (eds) 1980. *Thresholds in geomorphology.* London: George Allen & Unwin.

Cole, J. P. and C. A. M. King 1968. *Quantitative geography*. London: Wiley.

Coleman, A. 1952. Some aspects of the development of the lower Stour, Kent. *Proc. Geol Assoc.* **63**, 63–86.

Cooke, R. U. and J. C. Doornkamp 1974. *Geomorphology in environmental management*. Oxford: Oxford University Press.

Cooke, R. U. and D. K. C. Jones 1980. Suitable sites for cities. *Geogr. Mag.* **52**, 356–8.

Cooke, R. U. and A. Warren 1973. *Geomorphology in deserts*. London: Batsford.

Cotton, C. A. 1942a. *Landscape as developed by the processes of normal erosion*. 1st edn. Christchurch, New Zealand: Whitcombe & Tombs.

Cotton, C. A. 1942b. *Climatic accidents in landscape making*, 2nd edn. Christchurch, New Zealand: Whitcombe & Tombs.

Cotton, C. A. 1944. *Volcanoes as landscape forms*. Christchurch, New Zealand: Whitcombe & Tombs.

Craik, K. H. 1970. Environmental psychology. *New Directions in Psychology* **4**, 1–121.

Crickmay, C. H. 1933. The later stages in the cycle of erosion. *Geol Mag.* **70**, 337–47.

Cullingford, R. A., D. A. Davidson and J. Lewin (eds) 1980. *Timescales in geomorphology*. New York: Wiley.

Cvijic, J. 1918. L'hydrographie souterraine et l'évolution morphologique du karst. *Rev. Géogr. Alp.* **6**, 375–426.

Darwin, C. 1859. *On the origin of species*. London: Murray.

Davies, I. 1982. Safe houses in the Karakoram. *Geogr. Mag.* **54**, 30–9.

Davies, J. L. 1980. *Geographical variation in coastal development*, 2nd edn. London: Longman.

Davis, W. M. 1895. The development of certain English rivers. *Geogr. J.* **5**, 128–46.

Davis, W. M. 1899. The geographical cycle. *Geogr. J.* **14**, 481–504.

Davis, W. M. 1905. The geographical cycle in an arid climate. *J. Geol.* **13**, 381–407.

Davis, W. M. 1909. *Geographical essays*. Boston: Ginn; republished 1954, New York: Dover.

Davis, W. M. 1932. Piedmont beachlands and primarrümpfe. *Bull. Geol. Soc. Am.* **43**, 399.

Demek, J. (ed.) 1972. *Manual of detailed geomorphological mapping*. Prague: Academia.

Derbyshire. E. 1973. *Climatic geomorphology*. London: Macmillan.

Derbyshire, E., K. J. Gregory and J. R. Hails (eds) 1979. *Geomorphological processes*. Folkestone: Dawson.

Derbyshire, E. and C. H. B. Sperling 1981. The right materials for the job. *Geogr. Mag.* **53**, 455–7.

Detwyler, T. R. 1971. *Man's impact on environment*. New York: McGraw-Hill.

Diem, A. and J. Conway 1982. Swiss tame the Alpine snows. *Geogr. Mag.* **54**, 682–9.

Dixey, F. 1962. Applied geomorphology. *S. Afr. Geogr. J.* **44**, 3–24.

Doehring, D. O. (ed.) 1977. *Geomorphology in arid regions*. London: George Allen & Unwin.

Doornkamp, J. C. 1977. Rift valleys and recent tectonics. In *The unquiet landscape*, D. Brunsden and J. C. Doornkamp (eds), 21–6. Newton Abbot: David & Charles.

Doornkamp, J. C. and C. A. M. King 1971. *Numerical analysis in geomorphology*. London: Edward Arnold.

Douglas, I. 1967. Man, vegetation and the sediment yield of rivers. *Nature* **215**, 925–8.

Douglas, I. 1968. *Field methods of water hardness determination*. GeoAbstracts Technical Bulletin, no. 1. Norwich: GeoAbstracts.

Douglas, I. 1971. Dynamic equilibrium in applied geomorphology: two case-studies. *Earth Science J.* **5**, 29–35.

Douglas, I. 1977. *Humid landforms: an introduction to systematic geomorphology*. Cambridge, Mass.: MIT Press.

Dury, G. H. 1959. *The face of the Earth.* 1st edn. London: Penguin.

Dury, G. H. 1969. *Perspectives on geomorphic processes.* Assoc. Am. Geogr. Resource Pap., no. 3.

Dury, G. H. (ed.) 1970. *Rivers and river terraces.* London: Macmillan.

Dury, G. H. 1976. *The face of the Earth,* 2nd edn. London: Penguin.

Easterbrook, D. J. 1969. *Principles of geomorphology.* New York: McGraw-Hill.

Eiby, G. A. 1967. *Earthquakes.* London: Frederick Muller.

Embleton, C. 1972. *Glaciers and glacial erosion.* London: Macmillan.

Embleton, C., D. Brunsden and D. K. C. Jones (eds) 1978. *Geomorphology: present problems and future prospects.* Oxford: Oxford University Press.

Embleton, C. and C. A. M. King 1968. *Glacial and periglacial processes.* London: Edward Arnold.

Embleton, C. and C. A. M. King 1975a. *Glacial geomorphology.* London: Edward Arnold.

Embleton, C. and C. A. M. King 1975b. *Periglacial geomorphology.* London: Edward Arnold.

Embleton, C. and J. Thornes 1979. *Processes in geomorphology.* London: Edward Arnold.

Escritt, A. 1980. Unpredictable Krafla takes volcano watchers by surprise. *Geogr. Mag.* **53**, 1–12.

Evans, I. S. 1972. General geomorphometry, derivations of altitude and descriptive statistics. In *Spatial analysis in geomorphology,* R. J. Chorley (ed.) 17–90.

Everard, C. E. 1954. The Solent river. *Trans Inst. Br. Geogs* **20**, 41–58.

Fagg, C. C. 1923. The recession of the Chalk escarpment and the development of the dry chalk valley. *Proc. Croydon Nat. Hist. and Scient. Soc.* **9**, 93–112.

Fenneman, N. M. 1936. Cyclic and non-cyclic aspects of erosion. *Bull. Geol. Soc. Am.* **47**, 173–86.

Fines, K. D. 1968. Landscape evaluation: a research project in East Sussex. *Regional Studies* **2**, 41–55.

Finlayson, B. and I. Statham 1980. *Hillslope analysis.* London: Butterworth.

Flint, R. F. 1971. *Glacial and Quaternary geology.* New York: Wiley.

Ford, T. D. and C. H. D. Cullingford (eds) 1976. *The science of speleology.* New York: Academic Press.

Francis, P. 1976. *Volcanoes.* London: Penguin.

Francis, P. 1980. The blast that moved a mountain. *Geogr. Mag.* **53**, 729–38.

French, H. M. 1976. *The periglacial environment.* London: Longman.

Gaillard, D. B. W. 1904. *Wave action.* Washington DC: Corps of Engineers, US Army.

Gardner, J. 1977. *Physical geography.* New York: Harper & Row.

Garner, H. F. 1974. *The origin of landscapes: a synthesis of geomorphology.* Oxford: Oxford University Press.

Garwood, E. J. 1910. Features of Alpine scenery due to glacial protection. *Geogr. J.* **36**, 310–39.

Gass, I. G., P. J. Smith and R. C. L. Wilson (eds) 1972. *Understanding the earth,* 2nd edn. Horsham: Artemis Press.

Geer, G. de 1912. A geochronology of the last 12 000 years. C.R. XI Int. Geol. Congr. (Stockholm) **1**, 241–53.

Geikie, A. 1865. *The scenery of Scotland.* London.

George, F. H. 1967. The use of models in science. In *Models in geography,* R. J. Chorley and P. Haggett (eds), Ch. 2. London: Methuen.

Gerrard, J. 1981. *Soils and landforms.* London: George Allen & Unwin.

Gilbert, G. K. 1877. *Report on the geology of the Henry Mountains.* Washington: US Geol. Surv.

Gilbert, G. K. 1906. Crescentric gouges on glaciated surfaces. *Bull. Geol Soc. Am.* **17**, 303–13.

Gilbert, G. K. 1914. *The transportation of debris by running water.* US Geol. Surv. Prof. Pap., no. 86.

Gilman, K. and M. D. Newson 1983. *Soil pipes and pipeflow: a hydrological study in upland Wales.* Norwich: GeoBooks.

Glenn, L. C. 1911. Denudation and erosion in the southern Appalachian region and the Monongahela basin. US Geol. Surv. Prof. Pap., no. 72.

Goudie, A. 1977. *Environmental change.* Oxford: Oxford University Press.

Goudie, A. (ed.) 1981a. *Geomorphological techniques.* London: George Allen & Unwin.

Goudie, A. 1981b. Fearful landscape of the Karakoram. *Geogr. Mag.* **53**, 306–12.

Goudie, A., R. U. Cooke and I. Evans 1970. Experimental investigation of rock weathering by salts. *Area* 42–8.

Goudie, A. and A. Watson 1981. *Desert geomorphology.* London: Macmillan.

Goudie, A. and J. Wilkinson 1977. *The warm desert environment.* Cambridge: Cambridge University Press.

Gregory, K. J. (ed.) 1977. *River channel changes.* New York: Wiley.

Gregory, K. J. and D. E. Walling 1973. *Drainage basin form and process.* London: Edward Arnold.

Gregory, S. 1963. *Statistical methods and the geographer.* London: Longman.

Griggs, D. T. 1936. The factor of fatigue in rock exfoliation. *J. Geol.* **44**, 781–96.

Hails, J. R. (ed.) 1977. *Applied geomorphology.* Amsterdam: Elsevier.

Hails, J. and A. Carr (eds) 1975. *Nearshore sediment dynamics and sedimentation.* New York: Wiley.

Hallam, A. 1973. *A revolution in the earth sciences.* Oxford: Clarendon Press.

Hammond, R. and P. S. McCullagh 1978. *Quantitative techniques in geography*, 2nd edn. Oxford: Clarendon Press.

Hanwell, J. D. and M. D. Newson 1973. *Techniques in physical geography.* London: Macmillan.

Heather D. C. 1981. *Plate tectonics.* London: Edward Arnold.

High, C. and F. K. Hanna 1970. *A method for the direct measurement of erosion on rock surfaces.* GeoAbstracts Technical Bulletin, no. 5. Norwich: GeoAbstracts.

Hills, R. C. 1970. *The determination of the infiltration capacity of field soils using the cylinder infiltrometer.* GeoAbstracts Technical Bulletin, no. 3. Norwich: GeoAbstracts.

Hilton, K. 1979. *Process and pattern in physical geography.* London: University Tutorial Press.

Hjülstrom, F. 1935. Studies of the morphological activities of rivers as illustrated by the River Fyris. *Bull. Geol. Inst. Univ. Uppsala* **25**, 221–527.

Hobbs, W. H. 1910. The cycle of mountain glaciation. *Geogr. J.* **35**, 146–63.

Hodgson, J. M., J. A. Catt and A. H. Weir 1967. The origin and development of clay-with-flints and associated soil horizons on the South Downs. *J. Soil Sci.* **18**, 85–102.

Hodgson, J. M., J. H. Rayner and J. A. Catt 1974. The geomorphological significance of clay-with-flints on the South Downs. *Trans Inst. Br. Geogs* **61**, 119–29.

Högbom, B. 1914. Über die geologische Bedeutung des Frostes. *Bull. Geol. Instn Univ. Uppsala* **12**, 257–389.

Holmes, A. 1978. *Principles of physical geology*, 3rd edn. London: Nelson.

Horton, R. E. 1924. Discussion of the distribution of intense rainfall and some other factors in the design of storm-water drains. *Proc. Am. Soc. Civil Engrs* **50**, 660–7.

Horton, R. E. 1932. Drainage basin characteristics. *Trans Am. Geophys. Union* **13**, 350–61.

Horton, R. E. 1933. The role of infiltration in the hydrologic cycle. *Trans Am. Geophys Union* **14**, 446–60.

Horton, R. E. 1945. Erosional development of streams and their drainage basins: hydrophysical approach to quantitative morphology. *Geol Soc. Am. Bull.* **56**, 275–370.

Howe, G. M., H. O. Slaymaker and D. M. Harding 1966. Flood hazard in mid-Wales. *Nature* **212**, 584–5.

Howe, G. M., H. O. Slaymaker and D. M. Harding 1967. Some aspects of the flood hydrology of the upper catchments of the Severn and Wye. *Trans Inst. Br. Geogs* **41**, 33–58.

Jacks, G. V. and R. O. Whyte 1939. *The rape of the Earth – a world survey of soil erosion.* London: Faber.

James, P. A. 1971. *The measurement of soil frost-heave in the field.* GeoAbstracts Technical Bulletin, no. 8. Norwich: GeoAbstracts.

Jennings, J. N. 1971. *Karst.* Cambridge: Cambridge University Press.

Johnson, D. W. 1919. *Shore processes and shoreline development.* New York: Wiley.

Johnson, W. D. 1904. The profile of maturity in alpine glacial erosion. *J. Geol.* **12**, 569–78.

Jolliffe, I. P. and A. G. McLellan 1980. Prospectors for sand and gravel. *Geogr. Mag.* **52**, 615–17.

Jones, D. K. C. 1977. Man-made landforms. In *The unquiet landscape*, D. Brunsden and J. C. Doornkamp (eds), 149–55. Newton Abbot: David & Charles.

Jones, D. K. C. 1981. *South-east and southern England.* London: Methuen.

Jones, J. A. A. 1983. *The nature of soil piping: a review of research.* Norwich: GeoBooks.

Jones, O. T. 1951. The drainage system of Wales and the adjacent regions. *Q. J. Geol Soc.* **107**, 201–25.

Kendall, P. F. 1902. A system of glacier lakes in the Cleveland Hills. *Q. J. Geol Soc.* **58**, 471–571.

King, C. A. M. 1966. *Techniques in geomorphology.* London: Edward Arnold.

King, C. A. M. 1972. *Beaches and coasts*, 2nd edn. London: Edward Arnold.

King, C. A. M. 1974. Coasts. In *Geomorphology in environmental management*, R. U. Cooke and J. C. Doornkamp, 188–222. Oxford: Oxford University Press.

King, C. A. M. 1976. *Landforms and geomorphology. Concepts and history.* Stroudsburg, Pa: Dowden, Hutchinson & Ross.

King, C. A. M. 1980. *Physical geography.* Oxford: Blackwell.

King, L. C. 1950. The study of the world's plainlands. *Q. J. Geol Soc. Lond.* **106**, 101–31.

King, P. B. and S. A. Schumm (eds) 1980. *The physical geography (geomorphology) of William Morris Davis.* Norwich: GeoBooks.

Kirkby, M. J. and R. P. C. Morgan 1981. *Soil erosion.* New York: Wiley.

Knapp, B. J. 1973. *A system for the field measurement of soil water.* GeoAbstracts Technical Bulletin, no. 9. Norwich: GeoAbstracts.

Knapp, B. J. (ed.) 1981. *Practical foundations of physical geography.* London: George Allen & Unwin.

Komar, P. D. 1976. *Beach processes and sedimentation.* Englewood Cliffs, NJ: Prentice-Hall.

Kuhn, T. S. 1962. *The structure of scientific revolutions.* Chicago: University of Chicago Press.

Langbein, W. B. and L. B. Leopold 1964. Quasi-equilibrium states in channel morphology. *Am. J. Sci.* **262**, 782–94.

Langbein, W. B. and S. A. Schumm 1958. Yield of sediment in relation to mean annual precipitation. *Trans Am. Geophys. Union* **39**, 1076–84.

Lapparent, A. de 1907. *Leçons de géographie physique*, 3rd edn. Paris: Masson.

Latham, J. P. 1966. Remote sensing of the environment. *Geogr. Rev.* **56**, 288–91.

Leggett, R. F. 1973. *Cities and geology.* New York: McGraw-Hill.

Leopold, L. B. 1968. *Hydrology for urban land planning – a guidebook on the hydrologic effects of urban land use.* US Geol. Surv. Circular, no. 554.

Leopold, L. B. 1969a. *Quantitative comparison of some aesthetic factors among rivers.* US Geol. Surv. Circular, no. 620.

Leopold, L. B. 1969b. Landscape aesthetics. *Nat. Hist.* (Oct.), 35–46.

Leopold, L. B. and W. B. Langbein 1962. *The concept of entropy in landscape evolution.* US Geol. Surv. Prof. Pap., no. 500A.

Leopold, L. B. and T. Maddock 1953. *The hydraulic geometry of stream channels and some physiographic implications.* US Geol. Surv. Prof. Pap., no. 252.

Leopold, L. B. and M. G. Wolman 1957. *River channel patterns – braided, meandering and straight.* US Geol. Surv. Prof. Pap., no. 282B.

Leopold, L. B., M. G. Wolman and J. P. Miller 1964. *Fluvial processes in geomorphology.* San Francisco: W. H. Freeman.

Le Pichon, X., J. Farancheteau and J. Bonnin 1973. *Plate tectonics.* Amsterdam: Elsevier.

Lewis, W. V. 1931. The effect of wave incidence on the configuration of a shingle beach. *Geogr. J.* **78**, 129–48.

Lewis, W. V. and M. M. Miller 1955. Kaolin model glaciers. *J. Glaciology* **2**, 533–8.

Lillesand, T. M. and R. W. Kiefer 1979. *Remote sensing and image interpretation.* New York: Wiley.

Linsley, R. K., M. A. Kohler and J. H. L. Paulhus 1949. *Applied hydrology.* New York: McGraw Hill.

Linton, D. L. 1955. The problem of tors. *Geogr. J.* **121**, 478–87.

Linton, D. L. 1964. Tertiary landscape evolution. In *The British Isles*, J. W. Watson and J. B. Sissons (eds). London: Nelson.

Linton, D. L. 1968. The assessment of scenery as a natural resource. *Scott. Geog. Mag.* **84**, 218–38.

Mabbutt, J. A. 1977. *Desert landforms.* Cambridge, Mass.: MIT Press.

Macar, P. 1946. *Principes de géomorphologie normale, étude des formes du terrain des régions à climat humide.* Liege: Masson.

Macar, P. 1955. Appalachian and Ardennes levels of erosion compared. *J. Geol.* **63**, 253–67.

McCullagh, P. 1978. *Modern concepts in geomorphology.* Oxford: Oxford University Press.

Macdonald, G. A. 1972. *Volcanoes.* New York: Prentice-Hall.

McKee, E. D. 1980. *A study of global sand seas.* Tunbridge Wells: Castle House.

Mackin, J. H. 1948. Concept of the graded river. *Geol Soc. Am. Bull.* **59**, 463–512.

Manley, G. 1966. A new geography. *The Guardian* (17 Mar.).

Martonne, E. de 1913. Le climat facteur du relief. *Scientia*, 339–55.

Martonne, E. de 1935. *Traité de géographie physique.* Paris: Colin.

Martonne, E. de 1940. Problèmes morphologiques du Brésil tropical atlantique. *Ann. Géogr.* **49**, 1–27, 106–29.

Martonne, E. de 1946. Géographie zonale: la zone tropicale. *Ann. Géogr.* **55**, 1–18.

Melhorn, W. N. and R. C. Flemal (eds) 1981. *Theories of landform development.* London: George Allen & Unwin.

Mitchell, C. W. and D. Gavish 1980. Land on which battles are lost and won. *Geogr. Mag.* **52**, 838–40.

More, R. 1967. Hydrological models in geography. In *Models in geography*, R. J. Chorley and P. Haggett (eds), 145–85. London: Methuen.

Morgan, M. A. 1967. Hardware models in geography. In *Models in geography*, R. J. Chorley and P. Haggett (eds), Ch. 17. London: Methuen.

Morgan, R. and H. Scoging 1981. Soil erosion can be controlled. *Geogr. Mag.* **53**, 922–4.

Morisawa, M. 1968. *Streams: their dynamics and morphology.* New York: McGraw-Hill.

Morisawa, M. (ed.) 1981. *Quantitative geomorphology*. London: George Allen & Unwin.
Newson, M. D. 1975. *Flooding and flood hazard in the UK*. Oxford: Oxford University Press.

Oakeshott, G. B. 1976. *Volcanoes and earthquakes, geologic violence*. New York: McGraw-Hill.
Ollier, C. D. 1969. *Weathering*. Edinburgh: Oliver & Boyd.
Ollier, C. D. 1981. *Tectonics and landforms*. London: Longman.
Oxburgh, E. R. 1972. Plate tectonics. In *Understanding the earth*, I. G. Gass, P. J. Smith and P. C. L. Wilson (eds), 263–85. Horsham: Artemis Press.

Paterson, W. 1969. *The physics of glaciers*. Oxford: Pergamon.
Peltier, L. C. 1950. The geographic cycle in periglacial areas as it is related to climatic geomorphology. *Ann. Ass. Am. Geogr.* **40**, 214–36.
Penck, A. and E. Brückner 1909. *Die Alpen im Eiszeitalter*. Leipzig: Tauchnitz.
Penck, W. 1924. *Die morphologische Analyse: ein Kapitel der physikalischen Geologie*. Stuttgart: Englehorns. Translated by H. Czech and K. C. Boswell 1953. *Morphological analysis of land forms*. London: Macmillan.
Perry, A. H. 1981. *Environmental hazards in the British Isles*. London: George Allen & Unwin.
Phillips, A. W. 1970. *The use of the Woodhead sea bed drifter*. GeoAbstracts Technical Bulletin, no. 4. Norwich: GeoAbstracts.
Pitty, A. F. 1971. *Introduction to geomorphology*. London: Methuen.
Pitty, A. F. (ed.) 1979. *Geographical approaches to fluvial processes*. Norwich: GeoAbstracts.
Playfair, J. 1802. *Illustrations of the Huttonian theory of the Earth*. London: Cadell & Davies.
Powell, J. W. 1875. *Exploration of the Colorado river of the West and its tributaries*. Washington: Government Printing Office.
Price, R. J. 1972. *Glacial and fluvioglacial landforms*. London: Longman.
Price, R. J. and D. E. Sugden (compilers) 1972. *Polar geomorphology*. Inst. Br. Geogs, Spec. Pub., no. 4.

Ramsay, A. C. 1860. *The old glaciers of Switzerland and North Wales*. London.
Rice, R. J. 1977. *Fundamentals of geomorphology*. London: Longman.
Richthofen, F. von 1886. *Führer für Forschungsreisende*. Hannover.
Ritchie, W. 1981. Where the land meets the sea. *Geogr. Mag.* **53**, 772–4.
Rittmann, A. and L. Rittmann 1976. *Volcanoes*. London: Orbis.
Ruhe, R. V. 1975. *Geomorphology*. Boston, Mass.: Houghton Mifflin.
Runcorn, S. K. (ed.) 1962. *Continental drift, sea floor spreading and plate tectonics: implications to the earth sciences*. London: Academic Press.

Saussure, H. B. de 1779–96. *Voyages dans les Alpes*. Paris.
Sawyer, K. E. 1975. *Landscape studies*, 2nd edn. London: Edward Arnold.
Scheidegger, A. E. 1970. *Theoretical geomorphology*. Berlin: Springer-Verlag.
Scheuchzer, J. J. 1723. *Itinera per Helvetiae Regiones Alpinas*. Leyden.
Schimper, K. 1837. Über die Eiszeit. *Mem. Soc. Helv. Sci. Nat.* **5**, 38–51.
Schumm, S. A. 1963. *The disparity between present rates of denudation and orogeny*. US Geol. Surv. Prof. Pap., no. 454H.
Schumm, S. A. 1972. *River morphology*. Stroudsberg, Pa: Dowden, Hutchinson & Ross.
Schumm, S. A. 1977a. *The fluvial system*. New York: Wiley.
Schumm, S. A. (ed.) 1977b. *Drainage basin morphology*. Stroudsberg, Pa: Dowden, Hutchinson & Ross.
Schumm, S. A. and R. W. Lichty 1965. Time, space and causality in geomorphology. *Am. J. Sci.* **263**, 110–19.

Schumm, S. A. and M. P. Mosley 1973. *Slope morphology.* Stroudsberg, Pa: Dowden, Hutchinson & Ross.

Selby, M. J. 1982. *Hillslope materials and processes.* Oxford: Oxford University Press.

Shaw, D. P. 1983. A barrier to tame the Thames. *Geogr. Mag.* **55**, 129–31.

Sherlock, R. L. 1922. *Man as a geological agent.* London: Witherby.

Shotton, F. W. 1953. The Pleistocene deposits of the area between Coventry, Rugby and Leamington and their bearing upon the topographical development of the Midlands. *Phil Trans R. Soc. B* **237**, 209–60.

Simpson, R. B. 1966. Radar, geographical tool. *Ann. Assoc. Am. Geogs* **56**, 80–9.

Skempton, A. W. and J. N. Hutchinson 1969. *Stability of natural slopes and embankment foundations.* State of the Art Volume, 7th Int. Conf. Soil Mech. and Foundn Engng (Mexico), 291–340.

Small, R. J. 1978. *The study of landforms,* 2nd edn. Cambridge: Cambridge University Press.

Smith, D. I. and P. Stopp 1978. *The river basin: an introduction to the study of hydrology.* Cambridge: Cambridge University Press.

Sparks, B. W. 1949. The denudation chronology of the dip-slope of the South Downs. *Proc. Geol Assoc.* **60**, 165–215.

Sparks, B. W. 1971. *Rocks and relief.* London: Longman.

Sparks, B. W. 1972. *Geomorphology,* 2nd edn. London: Longman.

Sparks, B. W. and R. G. West 1972. *The Ice Age in Britain.* London: Methuen.

Statham, I. 1977. *Earth surface sediment transport.* Oxford: Oxford University Press.

Steers J. A. 1964. *The coastline of England and Wales.* Cambridge: Cambridge University Press.

Steers, J. A. 1980. *Coastal features of England and Wales: eight essays.* Cambridge: Oleander Press.

Stoddart, D. R. 1969. Climatic geomorphology: review and reassessment. *Prog. in Geog.* **1**, 160–222.

Strahler, A. N. 1950. Equilibrium theory of erosional slopes, approached by frequency distribution analysis. *Am. J. Sci.* **248**, 673–96 and 800–14.

Strahler, A. N. 1952. Dynamic basis of geomorphology. *Bull. Geol. Soc. Am.* **63**, 923–38.

Strahler, A. N. 1954. Statistical analysis in geomorphic research. *J. Geol.* **62**, 1–25.

Strahler, A. N. 1973. *Introduction to physical geography,* 3rd edn. New York: Wiley.

Strahler, A. N. 1975. *Physical geography,* 4th edn. New York: Wiley.

Strahler, A. N. 1976. *Principles of earth science.* New York: Harper & Row.

Strøm, K. M. 1949. The geomorphology of Norway. *Geogr. J.* **112**, 19–27.

Suess, E. 1888. *Das Antlitz der Erde.* Vienna: Tempsky.

Sugden, D. E. and B. John 1976. *Glaciers and landscape: a geomorphological approach.* London: Edward Arnold.

Surell, A. 1841. *Études sur les torrents des Hautes-Alpes.* Paris: l'Academie des Sciences.

Sweeting, M. M. 1972. *Karst landforms.* London: Macmillan.

Tank, R. W. 1973. *Focus on environmental geology.* Oxford: Oxford University Press.

Tarling, D. H. and M. P. Tarling 1973. *Continental drift.* London: Penguin.

Thomas, M. F. 1974. *Tropical geomorphology.* London: Macmillan.

Thomas, M. F. 1977. Landforms in equatorial rainforest areas. in *The unquiet landscape,* D. Brunsden and J. C. Doornkamp (eds), 137–42. Newton Abbot: David & Charles.

Thorarinsson, S. 1939. Observations on the drainage and rates of denudation in the Hoffellsjökull district. *Geog. Ann.* **21**, 189–215.

Thornbury, W. D. 1954. *Principles of geomorphology.* New York: Wiley.

Thornes, J. B. and D. Brunsden 1977. *Geomorphology and time.* London: Methuen.

Toulmin, S. and J. Goodfield 1965. *The discovery of time.* London: Hutchinson.

Townshend, J. R. G. (ed.) 1981. *Terrain analysis and remote sensing.* London: George Allen & Unwin.

Tricart, J. 1974. *Structural geomorphology.* London: Longman.

Tricart, J. and A. Cailleux 1972. *Introduction to climatic geomorphology.* London: Longman.

Twidale, C. R. 1971. *Structural landforms.* Cambridge, Mass.: MIT Press.

Twidale, C. R. 1976. *Analysis of landforms.* New York: Wiley.

Verstappen, H. Th. 1968. *Geomorphology and environment.* Inaugural address. Chair of Geomorphology, ITC Information Publication, Delft, now at Enschede, Holland.

Vine, F. J. and H. H. Hess 1970. Sea-floor spreading. In *The sea,* vol. 4. New York: Wiley Interscience.

Voight, B. 1977. *Rockslides and avalanches.* Amsterdam: Elsevier.

Waltham, A. C. 1978. *Catastrophe: the violent earth.* London: Macmillan.

Ward, R. C. 1978. *Floods.* London: Macmillan.

Warren, A. and M. Mainguet 1982. Defences against the aeolian threat. *Geogr. Mag.* **54,** 165–7.

Washburn, A. L. 1973. *Periglacial processes and environments.* London: Edward Arnold.

Washburn, L. 1979. *Geocryology: a survey of periglacial processes and environments.* London: Edward Arnold.

Wegener, A. 1912. Die Entstehung der Kontinente. *Petermanns Mitteilungen,* 185–95, 253–6, 305–9.

West, R. G. 1977. *Pleistocene geology and biology.* London: Longman.

Weyman, D. R. 1975. *Runoff processes and streamflow modelling.* Oxford: Oxford University Press.

Weyman, D. R. 1981. *Tectonic processes.* London: George Allen & Unwin.

Weyman, D. R. and V. Weyman 1977. *Landscape processes.* London: George Allen & Unwin.

Whalley, B. W. 1976. *Properties of materials and geomorphological explanation.* Oxford: Oxford University Press.

White, G. F. 1945. *Human adjustment to floods.* University of Chicago, Department of Geography Research Paper, no. 29.

White, G. F. (ed.) 1961. *Papers on flood problems.* University of Chicago, Department of Geography Research Paper, no. 70.

White, G. F. 1964. *Choice of adjustment to floods.* University of Chicago, Department of Geography Research Paper, no. 93.

White, G. F. (ed.) 1974. *Natural hazards.* Oxford: Oxford University Press.

White, G. F. *et al.* 1958. *Change in urban occupance of flood plains in the United States.* University of Chicago, Department of Geography Research Paper, no. 87.

Wiegel, R. L. (ed.) 1976. *Earthquake engineering.* Englewood Cliffs, NJ: Prentice-Hall.

Wilson, J. T. 1977. *Continents adrift and continents aground: readings from Scientific American.* San Francisco: W. H. Freeman.

Wolman, M. G. and J. P. Miller 1960. Magnitude and frequency of forces in geomorphic processes. *J. Geol.* **68,** 54–74.

Wood, R. M. and J. Hunt 1981. High road to lonely Shingshal. *Geogr. Mag.* **53,** 504–10.

Wooldridge, S. W. and D. L. Linton 1955. *Structure, surface and drainage in South-East England,* 2nd edn. London: George Philip.

Yatsu, E. 1966. *Rock control in geomorphology.* Tokyo: Sozosha.

Young, A. 1972. *Slopes.* London: Longman.

Zaruba, Q. and V. Mencl 1969. *Landslides and their control.* Amsterdam: Elsevier.

Subject index

224

Index of place names

Milton Keynes UK
Ingram Content Group UK Ltd.
UKHW040104071024
449327UK00019B/802